# Dynamic Properties of Polymer Materials and their Measurements

by
**Kartik Srinivas**
**Managing Principal**
**CAE & Material Testing Services, AdvanSES**
Phone: +91-9624447567; E-mail: info@advanses.com

Advanced Scientific and Engineering Services (AdvanSES)
212, Shukan Mall, Sabarmati-Gandhinagar Highway
Motera, Sabarmati, Ahmedabad 380005 India.
http://www.advanses.com

# To,

Dr. Ken Cornelius (Professor, Wright State University, Dayton, OH)

Bob Samples (Founder, Akron Rubber Development Lab., Akron, OH)

My Parents, Santha and M. S. Srinivasan

Subha and Advait

# Contents

**Dynamic Properties of Polymer Materials and their Measurements**     1

.1    Introduction . . . . . . . . . . . . . . . . . . . . . . . . . . . . . . . . . 1

.2    Natural and Synthetic Polymers . . . . . . . . . . . . . . . . . . . . . . 3

.3    Polymers . . . . . . . . . . . . . . . . . . . . . . . . . . . . . . . . . . . 6

.4    Testing of Polymers and Elastomers . . . . . . . . . . . . . . . . . . . . 8

       .4.1    Basic Concepts of Stress and Strain in a Material . . . . . . . . . 11

.5    Viscoelastic Properties . . . . . . . . . . . . . . . . . . . . . . . . . . . 12

       .5.1    The Storage Modulus Curve for Polymers . . . . . . . . . . . . . 13

.6    Dynamic Properties . . . . . . . . . . . . . . . . . . . . . . . . . . . . . 17

.7    Stress Relaxation . . . . . . . . . . . . . . . . . . . . . . . . . . . . . . 24

.8    Creep . . . . . . . . . . . . . . . . . . . . . . . . . . . . . . . . . . . . 26

.9    Linear Viscoelastic Behaviour . . . . . . . . . . . . . . . . . . . . . . . 28

       .9.1    Kelvin-Voigt Model . . . . . . . . . . . . . . . . . . . . . . . . . 31

       .9.2    Maxwell Model . . . . . . . . . . . . . . . . . . . . . . . . . . . 33

       .9.3    Standard Linear Zener Model . . . . . . . . . . . . . . . . . . . 34

.10 Effect of Fillers on Dynamic Properties . . . . . . . . . . . . . . . . . 35

.11 Applications of Dynamic Material Characterization . . . . . . . . . . . . 35

    .11.1 Development and Testing of Anti-Vibration Mounts . . . . . . . . . . 36

    .11.2 Material Characterization Testing under High Strain Rates . . . . . 40

    .11.3 Development and Failure Analysis of Rubber Rollers . . . . . . . . 42

    .11.4 Viscoelastic Analysis of Tires . . . . . . . . . . . . . . . . . . 43

    .11.5 Non-linear Viscoelastic Tire Simulation Using FEA . . . . . . . . . 44

.12 The Correlation Between Frequency and Temperature . . . . . . . . . . . 46

.13 Time-Temperature Superposition (TTS) . . . . . . . . . . . . . . . . . . 47

    .13.1 Calculation of shift factor . . . . . . . . . . . . . . . . . . . 51

    .13.2 Generalized Steps for TTS . . . . . . . . . . . . . . . . . . . . 52

.14 Instruments for Dynamic Testing of Polymers . . . . . . . . . . . . . . . 53

.15 ASTM D5992 and ISO 4664-1 . . . . . . . . . . . . . . . . . . . . . . . . 61

.16 References . . . . . . . . . . . . . . . . . . . . . . . . . . . . . . . 64

# Dynamic Properties of Polymer Materials and their Measurements

## .1 Introduction

Naturally occuring polymers like wood, rubber, silk, etc. have been used for a long time owing to benefits such as ease of availablility, environmental compatibility, and suitability for various applications. Natural rubber, a naturally occuring elastomer is made from latex extracted from the plant *Hevea brasiliensis*. Malaysia, India, Thailand and Indonesia are some of the largest rubber manufacturing countries. The latex plant grows in tropical climates and once the trees are 4 to 6 years old they can be harvested. Latex is tapped from the sap of the tree by making incisions deep enough to tap the latex carrying vessels. The tapped latex is collected and stored at a suitable location for further processing. Natural rubber in its raw state is made up of long chains of polymers, which in turn are made of small molecules called monomers. It is sticky, and deforms easily at room temperature and not particularly useful as an engineering material. The properties of rubber are made engineering friendly by the process of vulcanization. The credit for the discovery of vulcanization goes to Charles Goodyear. Vulcanization is the process of sulphurizing the gum rubber, where crosslinks are formed between the polymer chains and sulphur. Crosslinking makes the rubber elastic and is responsible for providing the rubber with hardness, stiffness and other engineering properties.

Synthetic rubber is made in a laboratory from petroleum by products. Fritz Hofmann is credited with the discovery of synthetic rubber in 1909. The development of synthetic rubber took place in the first decades of the twentieth century. The development of synthetic rubber gave rise to elastomer compounding as a science. Synthetic elastomers were developed to obtain characterisitics like high temperature resistance, marine compatibility, resistance to industrial oils etc. Synthesization of different types of new rubber materials in the past few decades has led to some novel elastomers and composites that have replaced the use of traditional metals in many applications. China is the largest manufacturer and consumer of synthetic rubber.

Steel, aluminum and other materials are combined with soft elastomers to design products that can provide specific deformation characterisitics. The elastomer provides the flexibility or compliance and steel provides the required directional stiffness. Applications as automotive bushings, bridge bearing pads, spherical laminated bearings etc., utilize the bimaterial design to provide the required deformation characteristics and shear strain capabilities. Tire design utilizes rubber so as to provide variable stiffness characteristics at different locations of the tire. The tread rubber provides frictional resistance combined with high strain capabilities. The side-wall rubber provides fatigue resistance combined with lower stiffness and the wedge rubber is made up of soft material to provide fatigue resistance from repeated shear straining. A generic sports utility vehicle tire is sometimes made up of upto twenty different kinds of rubbers to provide the required ride and deformation characteristics. A passenger car typically has about 650 rubber components.

Figure(.1) shows the typical stress-strain curves for three main types of polymeric materials. The first curve shows the behavior of a brittle polymer, where the failure is sudden as soon as the ultimate stress is reached. This class of materials show high hardness and behavior similar to ceramic materials. The second curve shows the behavior of a polymer similar to many metals, as can be observed there is a definite yield point and a hardening region before the ultimate failure. This polymer exhibits a plastic behavior beyond their elastic limits. The third curve shows the behavior of a typical elastomer. This curve shows that the material can handle large recoverable strains without undergoing permanent deformations. The characterisitic of the material to undergo repeated deformations without showing a *set* is termed as Hyperelasticity. The nonlinear nature of the curve shows the continuous change in modulus of elasticity with respect to the strain levels. The "Modulus" in elastomer parlance is the value of stress at a particular value of strain.

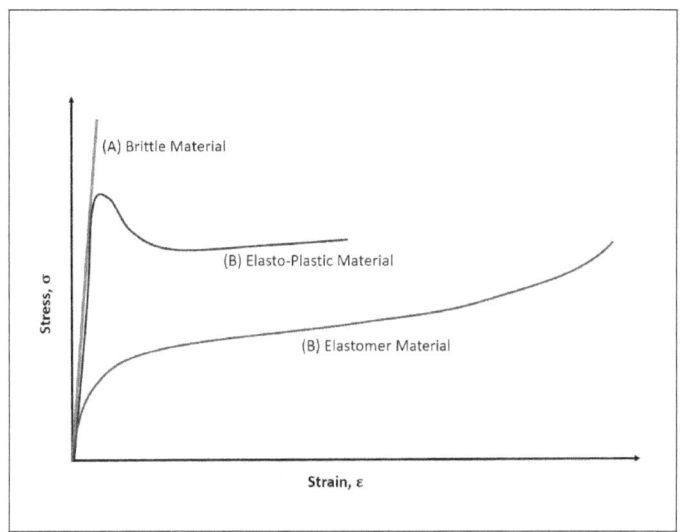

Figure .1: Stress-Strain Curves for Polymers

## .2 Natural and Synthetic Polymers

Polymers are classified into two broad categories, based on their behavior when exposed to high and low temperature, thermoset and thermoplastic. Thermoplastic polymers are manufactured in a two-step procedure. In the first step they are exposed to high temperature and they become soft and formable. The polymer melt are shaped in this softened state by pouring them into molds. When cooled below their melting point they rapidly again become rigid and form the shape of the mold. This type of polymers can be recycled easily and formed into new products over and over again. The changes seen in the thermoplastic material are purely physical and, with the reapplication of heat fully reversible.

As the name suggests, thermoset is a material that cures or hardens (sets) into a given shape, generally through the application of heat. Curing also known as vulcanizing is an irreversible chemical reaction in which permanent connections known as cross-links are made between the material's molecular chains. These intra-molecular cross-links as shownin Figure(.2) give the cured thermoset polymer a three-dimensional structure, as well as a higher degree of rigidity.

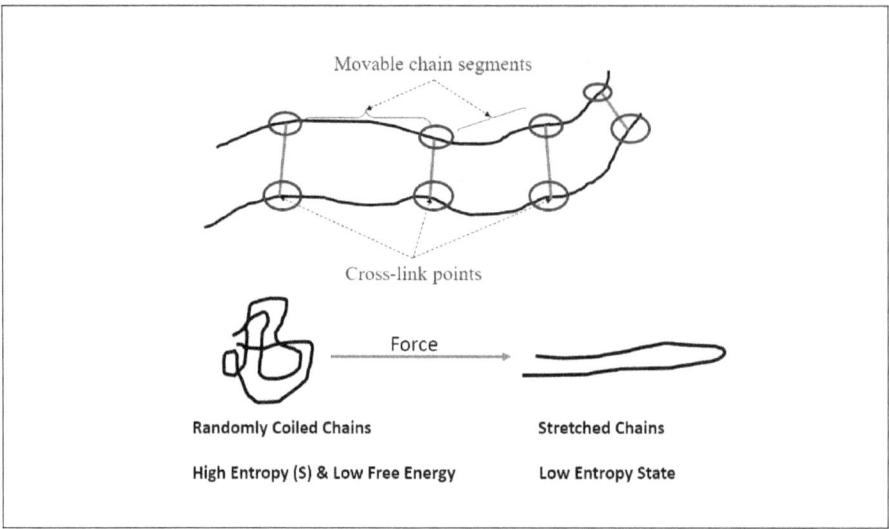

Figure .2: Thermoset Elastomers Chains and Motion

As can be seen an elastomer is made up of long chains. These chains under the effect of sulphur or any other vulcanizing agent forms cross-links giving it the necessary rigidity. The elongation of an elastomer is due to the elongation of these chains.

An elastomer product in undeformed or non- elongated state is at higher entropy and low free energy and when one stretches the elastomers there is a decrease in the entropy with an increase in free energy. The stretching of the elastomeric chains is akin to putting a pair of forks in a plate of spaghetti and moving the forks apart. The law of thermodynamics states that a material or a body always likes to be in a position of higher entropy rather than in a position of lower entropy, and this causes the material to come back to its original shape or position once the applied force is removed.

The amount and nature of cross-linking can be controlled at the compound formation level. Tight network of the molecular chains due to high crosslink levels restrict the motion of the chains preventing the network from elongation giving high rigidity. Very high levels of crosslinks are undesirable as they can lead to stress failures and too little crosslinks may not be strong enough to resist tensile failure. The optimum range of the cross-linking for practical use needs to be determined.

Curing changes the material forever. Thermoset polymers outperform other materials in a number of areas, including mechanical properties, chemical resistance, thermal stability, overall durability and provide components with high strain handling capability.

Synthetically manufactured elastomers have been developed seeking characteristics like high and low temperature durability, higher fracture toughness and longer fatigue life. Synthetic elastomers as SBR, Polybutadiene rubber, Isoprene, Neoprene, EPDM etc have been developed over a period of time to extract the best engineering properties from thermoset polymers.

Thermoplastic polymer resins consist of long polymer molecules which may or may not have side chains attached to them. The side chains are not linked to other polymer molecules as shown in Figure(.3). Thus there is an absence of cross-links in the thermoplastic structure. Thermoplastic resins in a granular form can be repeatedly melted or solidified by heating and cooling. Heat softens or melts the material so that it can be molded. Cooling in the mold solidifies the material into a given shape. There are two types of ther-

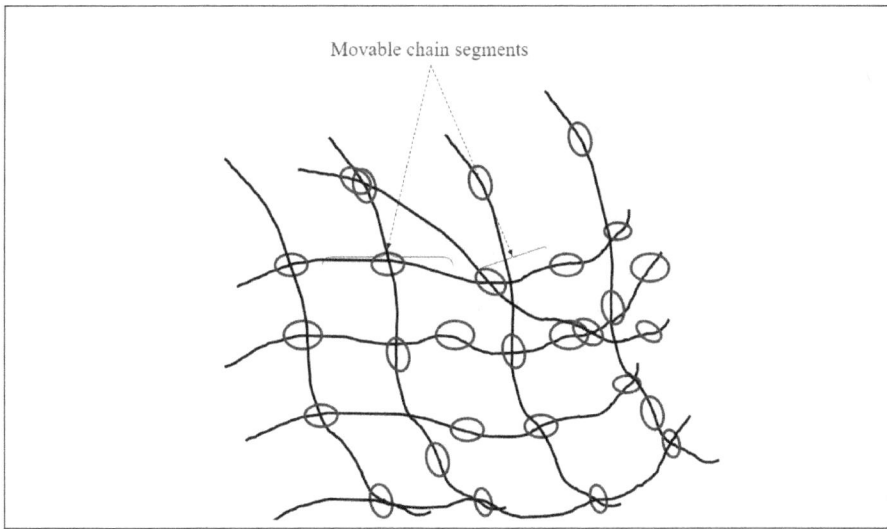

Figure .3: Chains in Thermoplastic Polymers

moplastic polymers, Crystalline and Amorphous. Following list enumerates the features and properties of both the polymer types.

**Crystalline Polymers**

1. Crystalline solids break along particular points and directions.

2. Crystalline solids have an ordered structural pattern of molecular chains.

3. Crystalline solids flow well at a higher temperature.

4. Reinforcement with fibers in cycstalline polymers increases the load-bearing capabilities.

5. Crystalline polymers tend to shrink more than amorphous.

6. The molecular structure of cystalline polymers makes them more suitable for opaque parts and components.

7. Examples: Polyethylene, Polypropylene, Nylon, Acetal, Polyethersulfone, etc.

**Amorphous Polymers**

1. Amorphous solids break into uneven parts with ragged edges.

2. Amorphous solids have a random orientation of molecules with no proper geometrical or pattern formation.

3. Amorphous solids do not flow as easily and can give problems in mold filling.

4. Examples: ABS, Polystyrene, Polycarbonate, etc.

Figure(.4) shows the general types and classification of polymers.

# .3 Polymers

Polymer materials in their basic form exhibit a range of characteristics and behavior from elastic solid to a viscous liquid. These behavior and properties depend on the temperature, frequency and time scale at which the material or the engineering component is analyzed. The viscous liquid polymer is defined as by having no definite shape and flow. Deformation under the effect of applied load is irreversible. Elastic materials such as steels and aluminium deform instantaneously under the application of load and return to the original state upon the removal of load, provided the applied load is within the yield limits of the

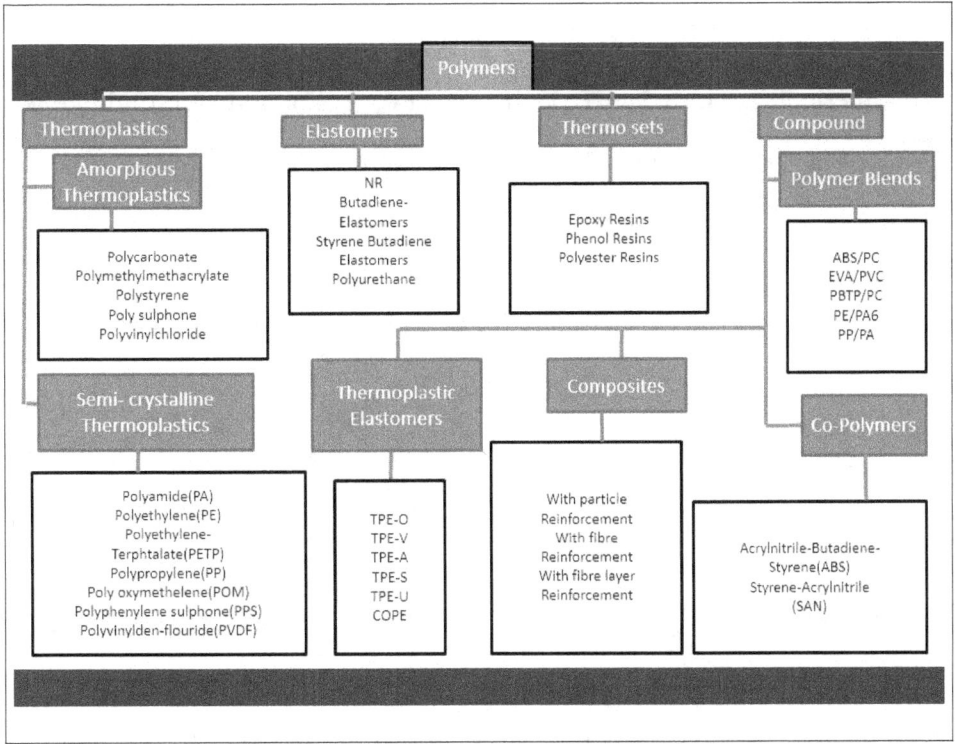

Figure .4: Types of Polymers and Their Classification

material. An elastic solid polymer is characterized by having a definite shape that deforms under external forces, storing this deformation energy and giving it back upon the removal of applied load.

**Ideal viscous and elastic materials**

1. Ideal elastic material:

    (a) The material obeys Hooke's law, wherein the stress is proportional to the applied strain.

    (b) Final deformation state is its original dimensions.

    (c) The deformation of the body is fully reversible.

2. Ideal viscous material:

    (a) The viscous stresses arising from its flow are linearly proportional to the local strain rate.

(b) Newtonian fluid.

Material behavior which combines both viscous liquid and solid like features is termed as Viscoelasticity. Viscoelastic materials exhibit a time-dependent behavior where the applied load does not cause an instantaneous deformation, but there is a time lag between the application of load and the resulting deformation. We also observe that in polymeric materials the resultant deformation also depends upon the speed of the applied load.

## .4 Testing of Polymers and Elastomers

The main objective of polymer testing is to characterize the material, make sure that it functions as it is designed, and check the quality for consistency and reliability. Polymer properties being dependent on time, temperature and deformation, most of the available tests in an engineer's arsenal cannot cover all of the characterisitics and properties. Due to these limitations all the test results, data and our inferences from these results can cover only a limited range and application. Overcoming this barrier involves testing the material and product in-situ under service conditions and test rigs assembled in the laboratory. Intimate knowledge and data on the interrelationship between the polymer compound recipe, the service conditions, the applied forces and boundary conditions and the effect of ageing on the material and product is necessary when it comes to improving the quality of the material and the product.

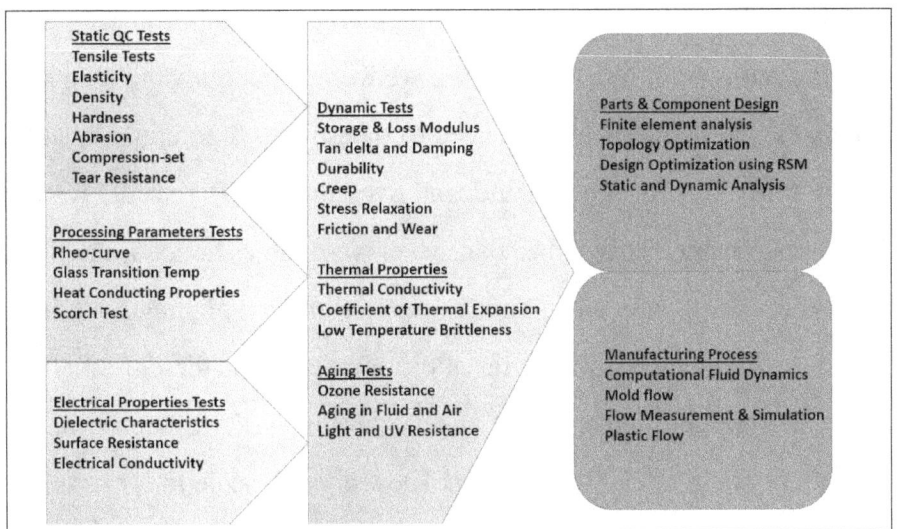

Figure .5: Elastomer and Polymer Testing

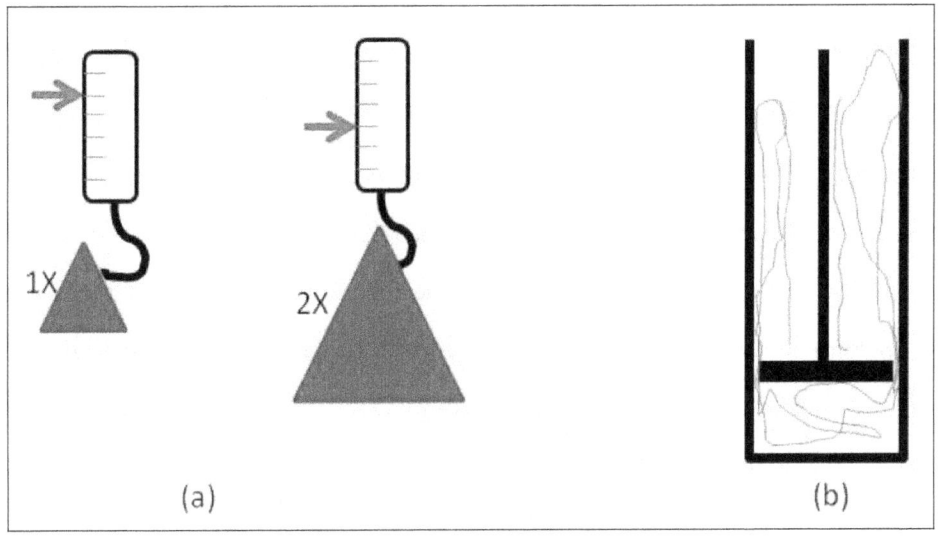

Figure .6: Linear Elastic and Non-linear Viscous Materials

Consider a spring based weighing scale as shown in Figure(.6). If we put a fixed load on the scale such as a weigh bar we get a fixed amount of deflection and the needle points to point 2. Using an exactly double weight the needle moves twice as far upto point 4 and comes to rest. When we remove the weights the needle returns back to 0 position or the no-load position. The energy we put into the spring based weighing scale to cause deflection is equal to the energy the spring fully gives up when the needle moves back to the initial zero

(0) position. This concept gives the spring perfectly linear properties and 100% efficiency. If we are to consider a situation where if we were to put our weight on the scale slowly such that it takes ten minutes before the needle comes to the 2 point mark, and if we did the same action in thirty seconds, one minute, and five minutes. The needle would still come to rest at the 2 point mark. This shows that when we double the load, the needle pointer moves twice as far as originally meaning the response of the weighing scale is linear. We can also observe that the weighing scale takes the same time or the deformation rate remains the same even if we apply the weight at different times. This shows that the deformation is independent of the rate of the loading. The weighing scale thus obeys Hooke's laws and is a perfect Hookean spring. The deformation energy is always stored as elastic energy and is fully given up upon removal of the load.

Consider a tall cylindrical jar filled with 5W40 engine oil as shown in Figure(.6b). At the bottom of the jar is a round disc. The disc is slightly smaller than the inside of the jar so that the oil can move around this disc. Applying an $X$ amount of force we can see that it takes 2 minute to extract the disc from the jar. To decrease the time taken by the disc to extract from the jar by 1 minute we see that we need to apply a higher force and to extract the disc in 30 seconds we need to apply an even higher force.

If the disc is moved from the bottom to the center of the jar and let it rest, it remains in the same spot and after some time under the force of gravity it starts sinking to the bottom. We can feel that by moving the disc through the engine oil we are applying force or putting work into the oil system by shearing it between the disc and the wall of the jar. The faster we try to move the disc, higher the force required. Unlike the independent nature of the force in the weighing scale, the amount of force required for the oil system now becomes rate dependent. e.g. It now depends upon how fast or slow we want to move the disc. The amount of force is not always proportional to the speed and now this becomes a non-linear relationship.

When we moved the disc from the bottom to the center of the jar, it stayed there and did not sink immediately to the bottom of the barrel. We put energy into the system but none of it was given back. All energy put into a viscous oil system is lost. The ideal viscous material is referred to as a Newtonian system.

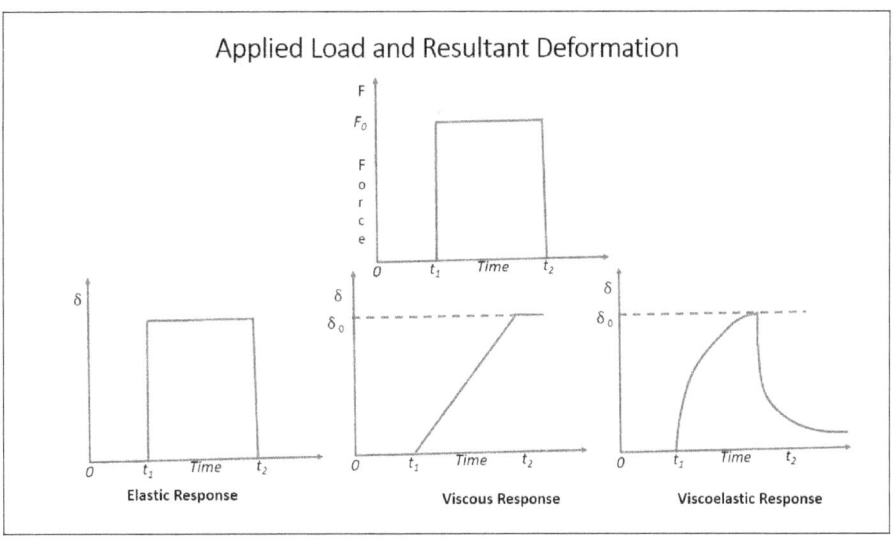

Figure .7: Responses of Elastic, Viscous and Viscoelastic Systems

Figure(.7) shows the applied load and the resulting deformations in a mechanical system. The load is applied in a step form from time $t_1$ to $t_2$. As can be seen the elastic response is linear and sudden as soon as the applied load is removed the defamation also goes out of the system. In a viscous response there is a certain time lag, the plot shows the time taken for the deformation to reach the final value. The viscoelastic response shows that it is a mixture or amalgamation of elastic and the viscous response. The time taken to reach the final deformation is nonlinear and varies over time. As soon as the load is removed one can see an appreciable decrease in the force decay. This phenomenon is purely a viscoelastic response of the system and is known as stress relaxation.

## .4.1 Basic Concepts of Stress and Strain in a Material

Load applied on a body results in the generation of stress in a body. The stress generated in the body is defined as

$$Stress = \frac{Force}{Area} \quad (.1)$$

The loading and the resultant stress describes the stress as tensile stress, compressive stres, and shear stress. Von-Mises stress, which is a function of the principal stresses, is generally taken as a representative average stress in the material to calculate endurance and failure limits.

$$Strain = \frac{deformation(\delta L)}{Original length(L)} \times 100 \quad (.2)$$

where δL is defined as final deformation - initial length $(L - L_0)$ Similar to the stress, the strain can also be defined as tensile strain, compressive strain, and shear strain.

The compliance of a material or system can be defined as;

$$Compliance = \frac{Strain}{Stress} \quad (.3)$$

The viscosity of a system, defined as measure of resistance to deformation by stress is given as,

$$Viscosity = \frac{Stress}{Applied Strain Rate} \quad (.4)$$

Plastics testing standards refer to a term called nominal strain which is defined differently depending on which test method is being used. For ASTM D638, nominal strain is defined as the strain measured from the crosshead displacement, not from the extensometer. This is because plastic does not break down homogenously, and strain is often focused on a disproportionately small part of the sample, a property called necking. For any materials that necks or has a yield point, percent elongation at break cannot be reported via the extensometer, as necking may occur outside of the extensometers gauge length. Therefore nominal strain must be used to report percent elongation at any points after yield. Using an extensometer for strain at break is only acceptable if the strain is homogenous throughout the specimen and doesnt exhibit necking or yield.

ASTM D412 is the test standard for tensile tests on rubber materials. Nominal strain is reported using a special long range extensometer capable of measure deformations of upto 1000 % strain. The standard has two methods, A and B, depending on the specimen geometry being tested. Method A refers to the use of dumbbell shaped specimens, while method B refers to ring shaped specimens.

## .5 Viscoelastic Properties

All materials exhibit viscoelastic characteristics upto come extent. In metallic materials such as steel, aluminum or even tungsten. These visccoleastic properties can be observed even at small strains at room temperature. The basic properties of these materials howevern does not deviate much from the behavior of linearly elastic. Natural and man made polymers like Natural rubber, synthetic rubber, plastics etc., show large viscoelastic effects at room and elevated temperatures due to their basic structure and their formation.

Some phenomena observed in viscoelastic materials are;

1. Elongation of a plastic thread under constant load, the strain increases with time (creep).

2. Decay in compression forced applied to a seal or o-ring in automotive applications. the stress decreases with time (stress relaxation).

3. Increasing stiffness of plastic materials with an increase in test speed.

4. Phase lag between the stress and strain when an automotive rubber component is subjected to sinusoidal loading.

5. A rubber ball dropped to the ground does not rise to the level it was dropped from.

Having observed complex failures in polymer parts during most of the last 5 decades, we know that design of polymer products can be successful only it we can incorporate their viscoelastic behavior of the material into the design stage itself. Testing and measurement of viscoelastic properties thus becomes critical to the design process itself.

## .5.1 The Storage Modulus Curve for Polymers

Polymer behaviour is both dependent on time and temperature

1. The Storage modulus curve (E') can be measured for both amorphous or partly crystalline polymers, as a function of time and temperature to understand long and short term characterisitics.

2. Polymer flow is affected by molecular motions and chain elongation and contraction.

3. For linear amorphous polymers five distinct regions of viscoelastic behaviour can be observed;

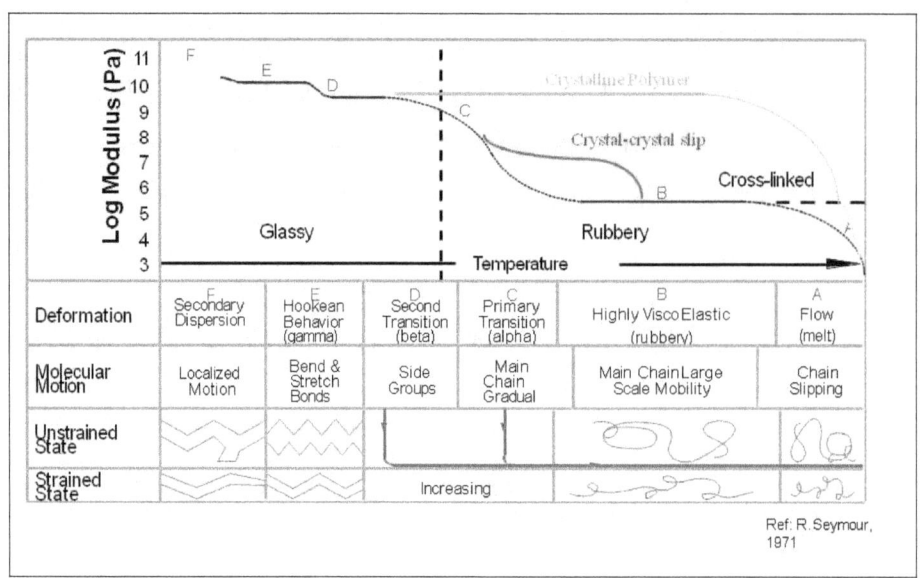

Figure .8: Different States and Modulus Regimes in Polymer Materials

1. Glassy Regime.

2. Glass transition ($T_g$) Regime.

3. Rubbery Plateau Region.

4. Rubbery-flow Region.

5. Liquid-flow Regime.

Figure .8 shows the regimes of a typical polymer material from the glassy to liquid state. The material is in the glassy state below the glass transition temperature, and the modulus of the material is as high as 3 GPa. The material practically behaves as a hard elastic material in this state. As the temperature increases the regime changes to glass transition state and the material modulus decreases from 3 GPa to approximately 10 MPa. Subsequently, the material enters the rubber regime or rubbery plateau and is effectively at room temperature. Most of our applications of polymer materials to engineering components fall within this plateau regime. Below the rubbery plateau is the viscous regime where the modulus of the material further decreases causing the material to flow like a liquid. Processing of plastic or rubber materials in a compression-transfer mold or injection mold takes place in this regime. Below the viscous regime is the region of decomposition and complete breakdown of the polymer material and its ingredients.

1. Polymer Properties and Characteristics in the Glassy region

    (a) The polymer is in a glassy and brittle state in this region and primarily behaves as a ceramic material.

    (b) Below $T_g$, the modulus value is approximately 1 GPa for many polymers.

    (c) Molecular motions are restricted to vibrations and short range rotational motions.

    (d) Chain elongations in this stage are not possible and lead to brittle failures.

2. Polymer Properties and Characteristics during the Glass Transition Regime

    (a) Test results show that the modulus drops by a factor of magnitude in a 20°C to 30°C range in this zone.

    (b) This stage is a defining characterisitic range for polymers and the $T_g$ value dictates how the materials will be used in engineering applications.

    (c) Precise testing instruments are used to measure and report the $T_g$ values.

3. Polymer Properties and Characteristics in the Rubbery plateau

    (a) Polymers are in their main application area in this regime. The rubbery plateau regime provides the necessary dynamic and static properties due to which elastomers are used in a wide array of applications.

    (b) Large strain causes entropic changes to the polymer material between the crosslinks.

    (c) Provides all the necessary properties at the crosslinks and chain level to undergo large deformations.

    (d) Modulus is inversely proportional to the molecular weight between entanglements.

4. Polymer Properties and Characteristics in the Rubbery-flow region

(a) The Rubbery-flow region provides the viscous flow properties to the solid form, enabling and aiding the manufacturing process of polymer products.

(b) The region combines the rubber elasticity and flow properties of a polymer material.

5. Polymer Properties and Characteristics in the Liquid flow region

    (a) Polymer flows readily under the effect of high temperature. This enables thermoplastic molding of plastic goods and products. Granules are melted in the heater and fed through the nozzle into a mold to enable production of polymer products.

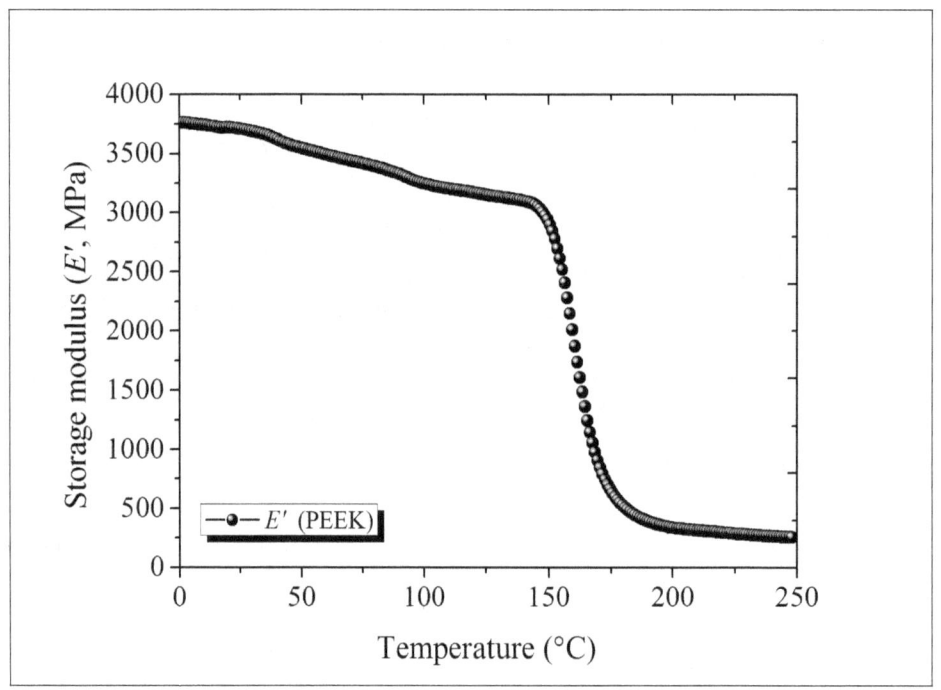

Figure .9: Storage Modulus Regime in PolyEther Ether Ketone (PEEK) Material

Figure(.9) shows the storage modulus plot for PolyEther Ether Ketone (PEEK) material from $0°C$ to $250°C$. The plot clearly shows the variation in the value of storage modulus with respect to temperature change. The molecular motion in the amorphous polymers being responsible for the changes. The glass transition is identified as the large drop in the modulus curve. DMA studies show that strain rate has a strong effect on the mechanical behavior of amorphous PEEK in rubbery state.

As the test frequency applied on a polymer sample is increased, the time required for molecular motion under cyclic deformation decreases. Instead of carrying out a DMA scan on temperature at constant frequency, a DMA scan can be carried under frequency sweep at constant temperature. If a sufficiently broad range of frequencies are scanned, all the transitions observed under a temperature scan can be found but with the observations being in reverse. The range of frequencies that materials can be tested at is limited because very low frequencies need a lot of time while very high frequencies are difficult to measure because of machine limitations. A practical approach to overcome these limitations is to carry out testing of viscoelastic behavior with a temperature scan than with a frequency scan.

## .6 Dynamic Properties

Static testing of materials as per ASTM D412, ASTM D638, ASTM D624 etc can be categorized as slow speed tests or static tests. The difference between a static test and dynamic test is not only simply based on the speed of the test but also on other test variables employed like forcing functions, displacement amplitudes, and strain cycles. The difference is also in the nature of the information we back out from the tests. When related to polymers and elastomers, the information from a conventional test is usually related to quality control aspect of the material or the product, while from dynamic tests we back out data regarding the functional performance of the material and the product.

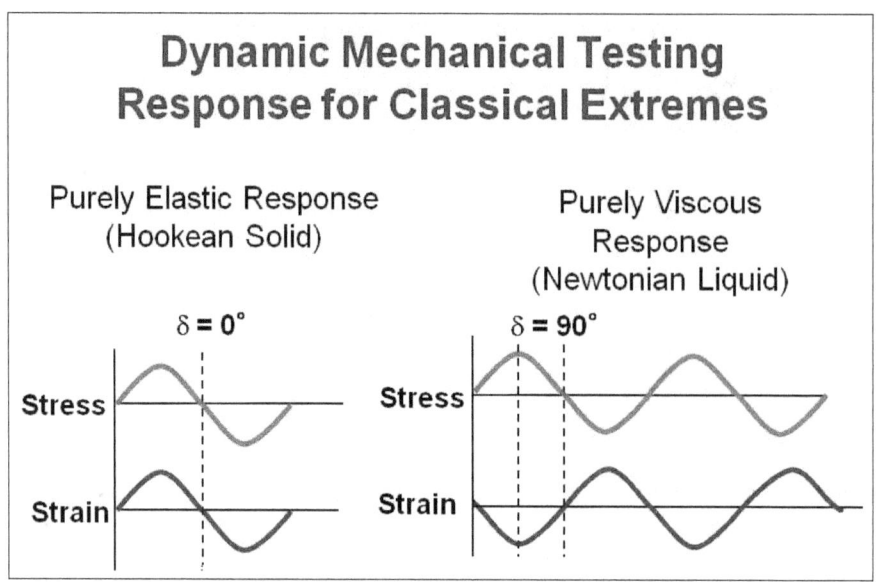

Figure .10: Purely Elastic and Purely Viscous Responses of Materials

Polymer materials are widely used in all kinds of engineering applications because of their superior performance in vibration isolation, impact resistance, rate dependency and time dependent properties. In some traditional applications they have consistently shown better performance combining with other materials like glass fibers etc.,and are now replacing metals and ceramics in such applications. The investigations of polymer properties in vibration, shock, impact and other viscoelastic phenomena is now considered critical, and understanding of dynamic mechanical behaviour of polymers becomes necessary and compulsory.

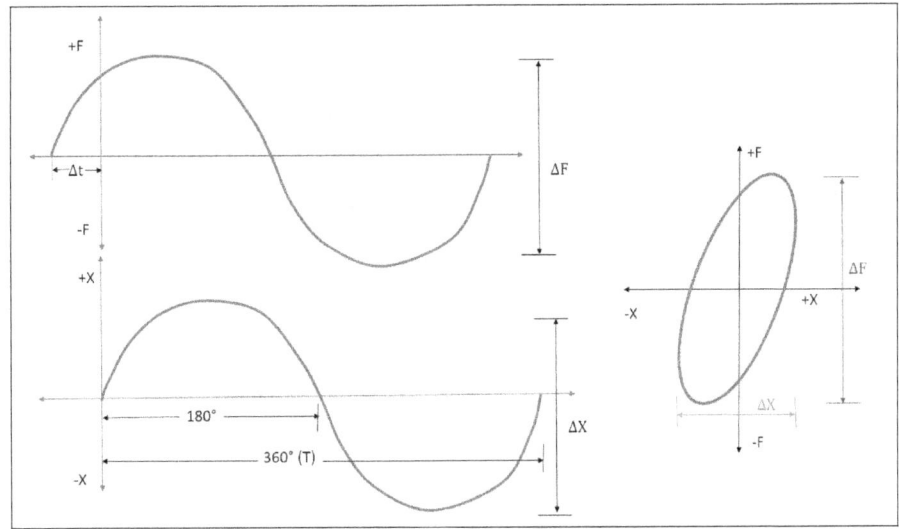

Figure .11: Dynamic Stiffness of a Polymer Material

Characterization of dynamic properties play an important part in comparing mechanical properties of different polymers for quality, performance prediction, failure analysis and new material qualification. Figures .10 and .12 show the responses of purely elastic, purely viscous and of a viscoelastic material. In the case of purely elastic, the stress and the strain (force and resultant deformation) are in perfect sync with each other, resulting in a phase angle of 0. For a purely viscous response the input and resultant deformation are out of phase by $90^o$. For a viscoleastic material the phase angle lies between 0 and $90^o$. Generally the measurements of viscoelastic materials are rerepresented as a complex modulus E* to capture both viscous and elastic behavior of the material. The stress is the sum of in-phase response and out-of-phase responses.

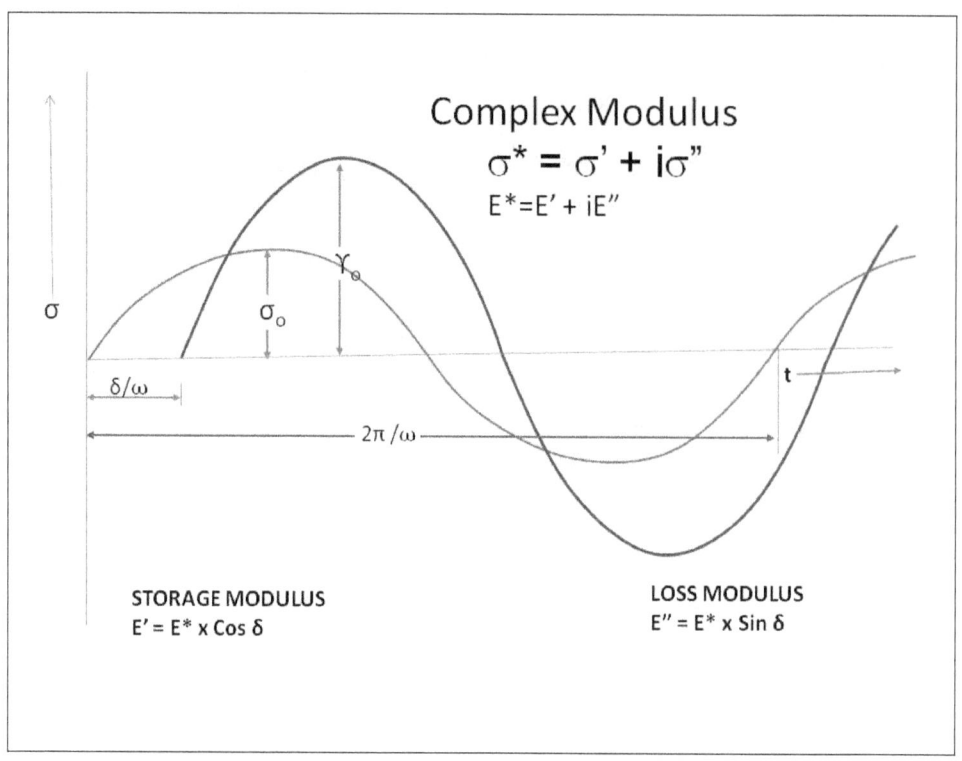

Figure .12: Viscoelastic Response of a Polymer Material

Figure .12 shows that the resultant strain lags the applied stress and can be written as,

$$\varepsilon(t) = \varepsilon_o Sin\omega t \qquad (.5)$$

the stress can be written as

$$\sigma(t) = \sigma_o Sin(\omega t + \delta) \qquad (.6)$$

where $\sigma_o$ is the stress amplitude and $\omega$ is the angular frequency. Expanding the equation,

$$\sigma(t) = \sigma_o Cos\delta Sin\omega t + \sigma_o Sin\delta Cos\omega t \qquad (.7)$$

The $\sigma_o Cos\delta$ term is in phase with the strain, while the term $\sigma_o Sin\delta$ is out of phase with the applied strain. The modulus $E'$ is in phase with the strain while, $E''$ is out of phase with the strain. The E' is termed as storage modulus, and E" is termed as the loss modulus.

$E' = \frac{\sigma_0}{\gamma_0} \cos\delta$

$E" = \frac{\sigma_0}{\gamma_0} \sin\delta$

$E' = \frac{In-phase\ stress}{Maximum\ strain} =\}$Storage Modulus

The storage modulus defines the elastic modulus part, where the deformations are fully recoverable at the end of the displacement cycle.

$$E" = \frac{Out\ of\ phase\ stress}{Maximum\ strain} = \}Loss\ Modulus$$

This shows that the strain energy associated with the in-phase stress and strain is reversible, i.e. that energy which is stored in the material during a loading cycle can be recovered without loss once the force is unloaded (Hyperelasticity). While the energy lost by the material due to the out-of-phase component is converted irreversibly to heat. The in-phase stress and strain results in completely recoverable elastic energy while the out-of-phase stress and strain results in the permanent dissipated energy termed as loss. Both the storage modulus and the loss modulus are functions of the applied frequency $\omega$.

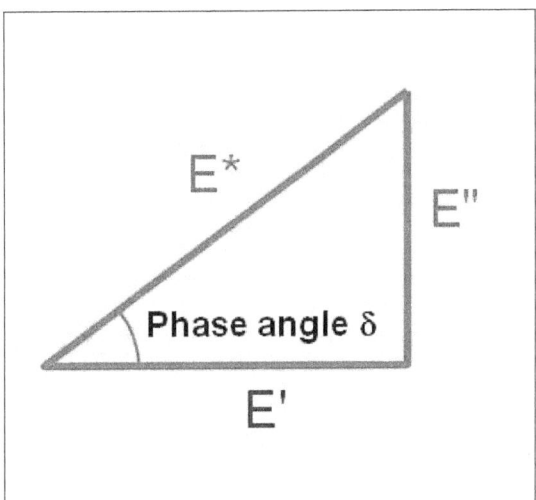

Figure .13: Concept of Complex, Loss and Storage Modulus

Figure(.13) shows the graphical representation of the loss modulus, storage and complex modulus. The loss tangent or the damping coefficient is defined as

$$tan\delta = \frac{E"}{E'} = \frac{Loss.Modulus}{Storage.Modulus} \tag{.8}$$

Figure(.14) shows the concept of loss and storage modulus from the example of a rubber ball. The ball when bounced off the ground from a height does not bounce to the level from where it was dropped but bounces only to a lower level. This loss of height can be defined as the energy that has been lost in energy disspation due to deformation of the rubber.

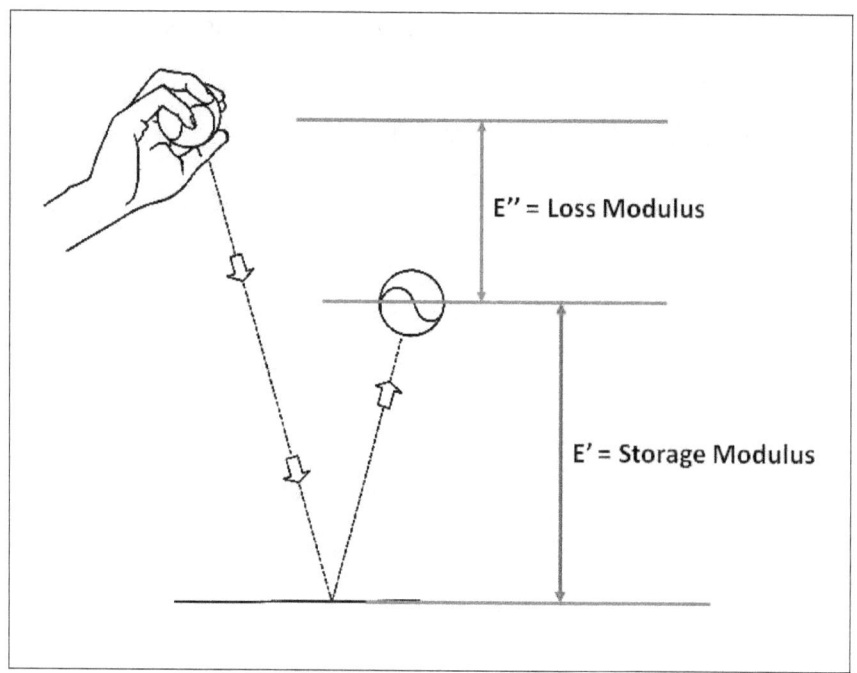

Figure .14: Loss and Storage Modulus in Elastomers

Polymer materials exhibit strong time and rate dependent properties. Studying these time and rate dependent properties is necessary to predict the performance of engineering components. To accurately characterize the viscoelastic properties of polymers, the generally used technique is that the material is sinusoidally deformed and the resulting stress is recorded. The deformation can be applied in tension, compression or shear mode depending on the geometric design or the application condition of the engineering component.

| Material | Temperature | Frequency | Tan δ |
|---|---|---|---|
| Sapphire | 4.2K | 30 Khz | $2.5 \times 10^{-10}$ |
| Sapphire | RT | 30 Khz | $5.0 \times 10^{-9}$ |
| Silicone | RT | 20 Khz | $3.0 \times 10^{-8}$ |
| Quartz | RT | 1 Mhz | approx. $10^{-7}$ |
| Aluminum | RT | 20 Khz | approx. $10^{-5}$ |
| Cu-31%Zn | RT | 6 Khz | $9.0 \times 10^{-5}$ |
| Steel | RT | 1 Hz | 0.0005 |
| Aluminum | RT | 1 Hz | 0.001 |
| Fe-0.6%V | 33C | 0.95 Hz | 0.0016 |
| Basalt | RT | 0.001-0.5 Hz | 0.0017 |
| Granite | RT | 0.001-0.5 Hz | 0.0031 |
| Glass | RT | 1 Hz | 0.0043 |
| Wood | RT | 1 Hz | 0.02 |
| Bone | 37C | 1-100 Hz | 0.01 |
| Lead | RT | 1-15 Hz | 0.001 |
| PMMA | RT | 1 Hz | 0.029 |
| Natural Rubber | RT | 5 Hz | 0.09 |

Table .1: Loss Tangents of Common Materials

Phase angle tan δ is associated with the degree of viscoelsticity of the sample and is a measure of damping provided by the material. A low value in tan δ indicates a higher degree of viscoelasticity (more elastic and solid like). The phase angle delta (δ) can be used to describe the properties of a sample.

**δ = 90 → G\*= G" and G'= 0 → viscous sample**

**δ = 0 → G\*= G' and G"= 0 → elastic sample**

**0 < δ < 90 → viscoelastic sample**

**δ > 45 → G" > G' → semi liquid sample**

**δ < 45 → G' > G" → semi solid sample**

The storage and loss modulus essentially represent the energy storage capacity and energy dissipation capacity of a polymer material. The extent and the amount to which the polymer chains can store and dissipate energy depends on the rate with which the chains and the crosslinks can change or alter their relative positions and undergo deformation with respect to time and frequency.

At lower frequencies and smaller amplitudes the chains deform less, and at higher frequencies and larger amplitudes the chains tend to deform more. These deformation characteristics of the polymer control the storage and loss modulli values of the material.

## .7 Stress Relaxation

Stress relaxation is a viscoelastic property of an elastomeric material. In a stress relaxation experiment, the sample is rapidly stretched or compressed to a predefined strain and held constant. The stress is then recorded as a function of time. Creep experiments are carried out in a similar manner but instead of the application of a constant strain, a constant load or stress is applied and the deformations or the resultant strain is studied as a function of time. The stress relaxation modulus may be define as

$$E_{rel.}(t) = \sigma(t)/\varepsilon_0 \tag{.9}$$

If there is no viscous flow in the material, the stress decays to a finite value for polymeric materials. For amorphous linear polymers at high temperatures, the stress may eventually decay to zero. For a linear viscoelastic solid, the instantaneous stress will be proportional to applied strain and will always decrease with time.

The molecular causes of stress relaxation and creep can be classified to be based on five different processes.

1. *Chain Scission*: The decrease in the measured stress over time is shown in Figure(.15) where 3 chains initially bear the load but subsequently one of the chains degrades and breaks.

2. *Bond Interchange*: In this particular type of material degradation process, the chain portions reorient themselves with respect to their partners causing a decrease in

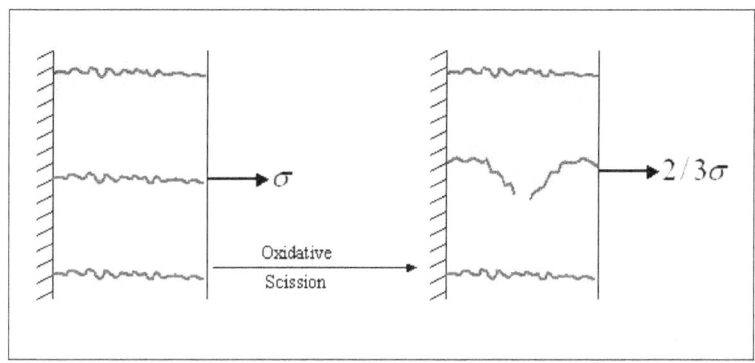

Figure .15: Chain Scission in an Elastomeric Material

stress.

3. *Viscous Flow*: This occurs basically due to the slipping of linear chains one over the other. It is particularly responsible for viscous flow in pipes and elongation flow under stress.

4. *Thirion Relaxation*: This is a reversible relaxation of the physical crosslinks or the entanglements in elastomeric networks. Generally an elastomeric network will instantaneously relax by about 5% through this mechanism.

5. *Molecular Relaxation*: Molecular relaxation occurs especially near $T_g$(Glass Transition Temperature). The molecular chains generally tend to relax near the $T_g$.

The related effect of stress relaxation in sealing applications is the reduction in stress (i.e. sealing force) under constant strain (compression) conditions. In many applications when sealing force decreases below a critical level leakage is likely to occur as the seal geometry loses contact between the mating surfaces. Stress relaxation and compression set effects are undesirable and are of key importance in determining sealing performance. For amorphous linear polymers at high temperatures, the stress may eventually decay to zero. If there is no viscous flow in the material, the stress decays to a finite non-zero value.

Stress relaxation as discussed above is caused by a combination of physical relaxation, and chemical degradation. Physical relaxation is dependent on elastomer material properties. It increases with temperature, is dominant over short timescales during the initial time period of the experiment and is fully recoverable. Chemical degradation is highly

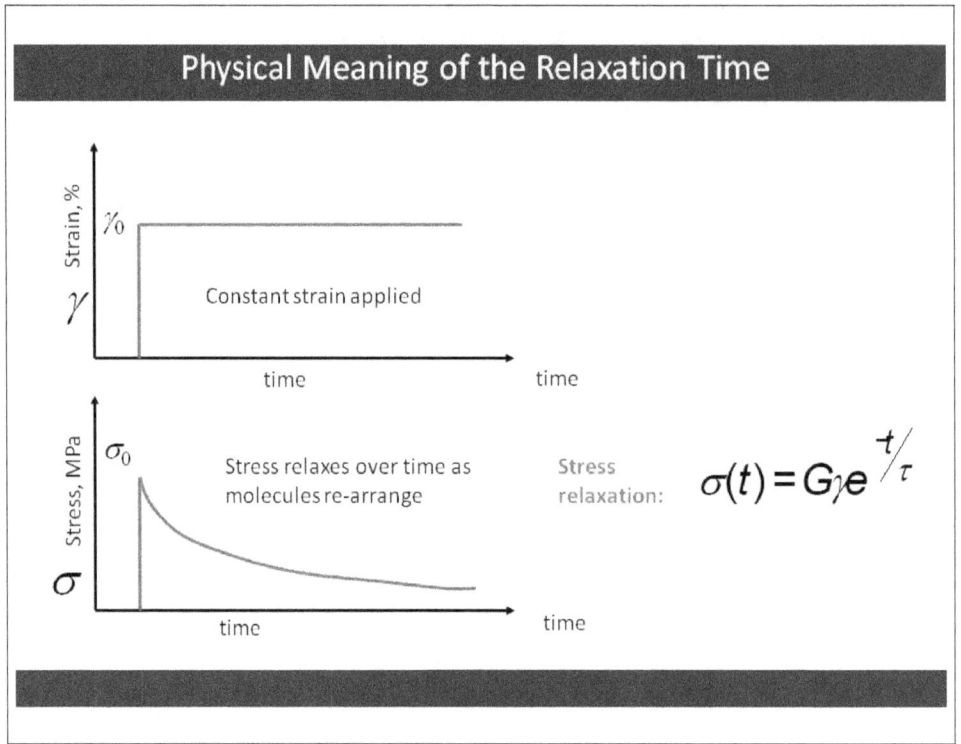

Figure .16: Physical Meaning of the Relaxation Time

dependent on temperature and on the aggressive chemical action of fluids and application environments and is applicable over longer time frames in an experiment. ASTM D6147 and ISO 3384 with methods A and B are the applicable standards for stress relaxation experiments. Method A refers to the continuous method of measuring the force over time, while method B refers to the intermittent method of measuring the force over time. Wykeham-Farrrance test apparatus is the preferred method to measure the force decay over time in method B. Fully automated test apparatus available in our laboratory at AdvanSES is capable of carrying out tests as per method A.

# .8 Creep

Creep is an increase in plastic strain under constant stress. Creep is an increased tendency of a solid material to move slowly or deform continuously under the influence of mechanical stresses. In other words it tends towards high strain and plastic deformation with no change in stress. Figure(.17) shows a the stress and strain curves for a part undergoing creep. The material is stressed with an applied force. Creep tends to occur as a result of long-term exposure to high levels of stress that are still below the yield strength of the material. Over time, the force and stress do not change, although the shape of the part continuously deforms. When unloaded, there is additional permanent set. Old PVC pipes for electrical installations sag at the center when simply supported at the ends. This is an example of creep under the constant force of gravity. Creep in polymers at low strains (1 percent) is essentially recoverable after unloading.

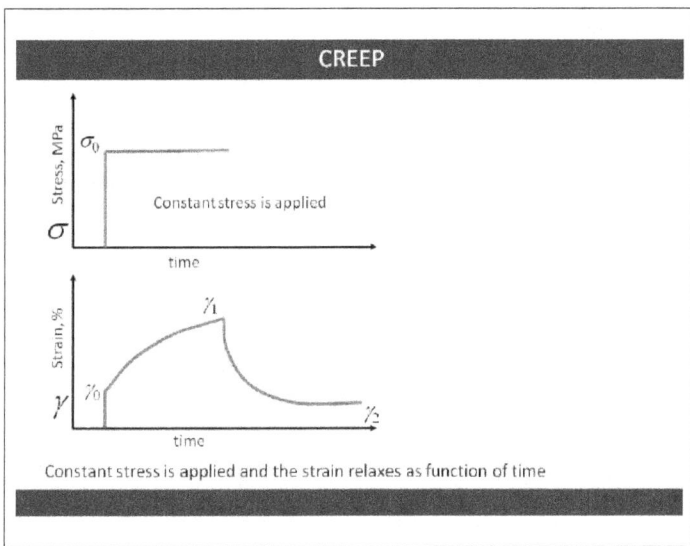

Figure .17: Physical Meaning of Creep

$\gamma_1$ is the immediate elastic deformation taking place under the effect of applied load. $\gamma_2$ is the elastic deformation that takes place over a period of time. $\gamma_3$ is the flow of the material or the non-recoverable creep that remains once the loading on the material is removed. The creep compliance may be define as

$$C_{comp.}(t) = \varepsilon_t/\sigma(0) \qquad (.10)$$

Creep is thus a time- dependent deformation under an externally applied load. It generally occurs at high temperature(thermal creep), but can also happen at room temperature in certain materials albeit at much slower rate. As a result of Creep the material undergoes a time dependent increase in length, which could be dangerous while occuring in engineering components under service loads and boundary conditions.

ASTM F38 is the standard test method for Creep experiments on a gasket material.

## .9 Linear Viscoelastic Behaviour

Polymer materials by their very nature exhibit linear and nonlinear viscoelasticity. Linear viscoelastic behavior is exhibited by a material that is subjected to a deformation that is either very small or very slow on the application time scale. The nature of the deformation can be predicted using a linear theoretical expression. When the polymers are subjected to a rapid rate of flow and deformation or sudden temperature cycling to the extremes, this results in deformation patterns that cannot be predicted by linear expressions or closed form solutions, and the material behavior is said to exhibit nonlinearity.

The mathematical development of modeling linear elastic behavior can be carried out using two (2) basic approaches 1) Phenomenological models and 2) Boltzmann Superposition Principle

1. **Phenomenological models**

    (a) Phenomenological models have no direct relationship with chemical composition or molecular structure.

    (b) Phenomenological models enable response to a complicated loading pattern to be deduced from a single creep or stress-relaxation plot.

    (c) The validity of phenomenological models depend upon the assumptions of linear viscoelasticity.

    (d) The key assumption with phenomenological models is that the total deformation can be considered as sum of independent elastic and viscous components.

2. **Boltzmann Superposition Principle**

    (a) The Boltzmann superposition principle is starting point for a theory of linear viscoelastic behavior. It relates the stress to the strain by a linear differential equation.

(b) The differential equation can be expresses as A σ = Bε, where A, B are linear differential operators.

(c) Each loading step during the deformation pattern makes an independent contribution to final deformation, and the total deformation is the sum of all deformation contributions.

(d) This is equivalent to describing the viscoelastic behaviour by mechanical models constructed of mechanical components like dashpots and springs.

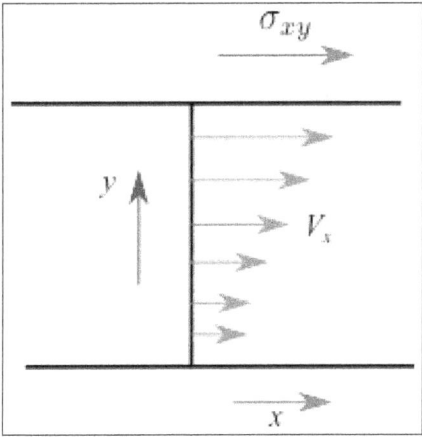

Figure .18: Viscous Flow Regime

Newtons law of viscosity defines viscosity by stating that stress is proportional to the velocity gradient in the liquid,

$$\sigma = \eta \frac{\partial V}{\partial y} \qquad (.11)$$

where $V$ is the velocity and $y$ is the direction of the velocity gradient. For a velocity gradient in the xy plane

$$\sigma_{xy} = \eta \left( \frac{\partial V_x}{\partial y} + \frac{\partial V_y}{\partial x} \right) \qquad (.12)$$

where $\frac{\partial V_x}{\partial y}$ and $\frac{\partial V_y}{\partial x}$ are the velocity gradients in the x and y directions.

As $V_x = \frac{\partial u}{\partial t}$ and $V_x = \frac{\partial v}{\partial t}$ with u and v as the displacements in the x and y directions, we get,

$$\sigma_{xy} = \eta \left\{ \frac{\partial}{\partial y}\left(\frac{\partial u}{\partial t}\right) + \frac{\partial}{\partial x}\left(\frac{\partial v}{\partial t}\right) \right\}$$
$$= \eta \frac{\partial}{\partial t}\left\{\frac{\partial u}{\partial y} + \frac{\partial v}{\partial x}\right\} \quad (.13)$$
$$= \eta \frac{\partial e_{xy}}{\partial t}$$

It can be seen that the shear stress $\sigma_{xy}$ is directly proportional to the rate of change of shear strain with time. This formulation brings out the analogy between Hooke's law for elastic solids and Newton's law for viscous liquids. For elastic solids the stress is linearly related to the strain but in the latter the stress is linearly related to the rate of change of strain or strain rate. Hooke's law describes the behaviour of a linear elastic solid and Newtons law that of a linear viscous liquid. A simple constitutive relation for the behaviour of a linear viscoelastic solid is obtained by combining these two laws: 1. For elastic behaviour $(\sigma_{xy})_E = Ge_{xy}$, where G is the shear modulus. 2. For viscous behavior $(\sigma_{xy})_V = \eta\left(\frac{\partial e_{xy}}{\partial t}\right)$. A simple possible formulation of linear viscoelastic behaviour combines these equations, making the assumption that the shear stresses related to strain and strain rate are

$$\sigma_{xy} = (\sigma_{xy})_E + (\sigma_{xy})_V = Ge_{xy} + \eta \frac{\partial e_{xy}}{\partial t} \quad (.14)$$

The equation represents one of the simple models for linear viscoelastic behaviour. For elastic solids Hooke's law is valid only at small strains, and Newton's law of viscosity is restricted to relatively low flow rates, as only when the stress is proportional either to the strain or the strain rate is analysis of the deformation feasible in simple form. A comparable limitation holds for viscoelastic materials. Quantitative predictions are possible only in the case of linear viscoelasticity, for which the results of changing stresses or strains are simply additive in nature.

A classical approach to the description of the linear viscoelastic behavior of real materials which exhibit combined viscous and elastic properties is based upon an analogy with the response of combinations of certain mechanical elements (a spring for elasticity, and a dashpot for viscosity). Springs and dashpots constitute the building blocks of model analyses in viscoelasticity. Springs and dashpots connected to one another in various forms are used to construct empirical viscoelastic models. Springs are used to account for the elastic solid behavior and dashpots are used to describe the viscous fluid behavior. These models are generalized ideal and hypothetical, however they are useful for representing the approximate behavior of real world materials.

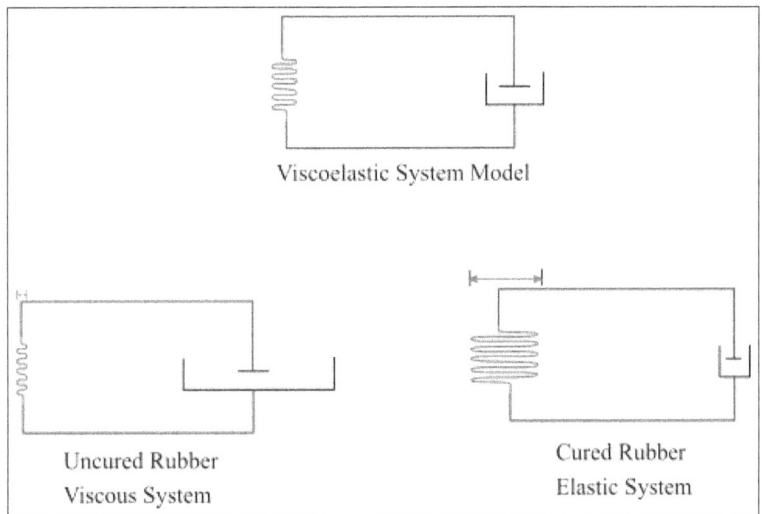

Figure .19: Viscous and Elastic Phases making up the Rubber Viscoelastic System

To summarize the response of the phases; The elastic response is linear and all the work input into the system is given up upon removal of applied load. The viscous response on the other hand is non-linear with the force and the deformation dependent upon the rate of the loading and the time. Joining the two phases; we now have a visco-elastic material that is graphically represented by a Hookean spring in parallel with a Newtonian dashpot, defined as the Kelvin-Voigt model as shown in Figure(.19). When the viscous and elastic components are joined, we can simultaneously put energy into both of them, but the response of each phase is almost on opposites. In uncured rubber, the viscous phase dominates as evidenced by the flow of the material in the mold. As the rubber is cured during vulcanization the elastic phase slowly begins to emerge and starts dominating the deformation characteristics.

## .9.1 Kelvin-Voigt Model

In this model, the total applied stress is subdivided into the stress borne by the spring and the stress borne by the dashpot; both elements undergo the same deformation. The elastic deformation of the spring depends on the viscosity of the liquid in the dashpot. Initial deformation of the Kelvin-Voigt body cannot occur instantaneously, because the dashpot requires some time to respond to the applied stress. Notably, if there is a near zero value of elastic modulus (G) for the spring, this body essentially becomes the simple linear viscous element. Hooke's law for an elastic solid states that stress is linearly related to strain,

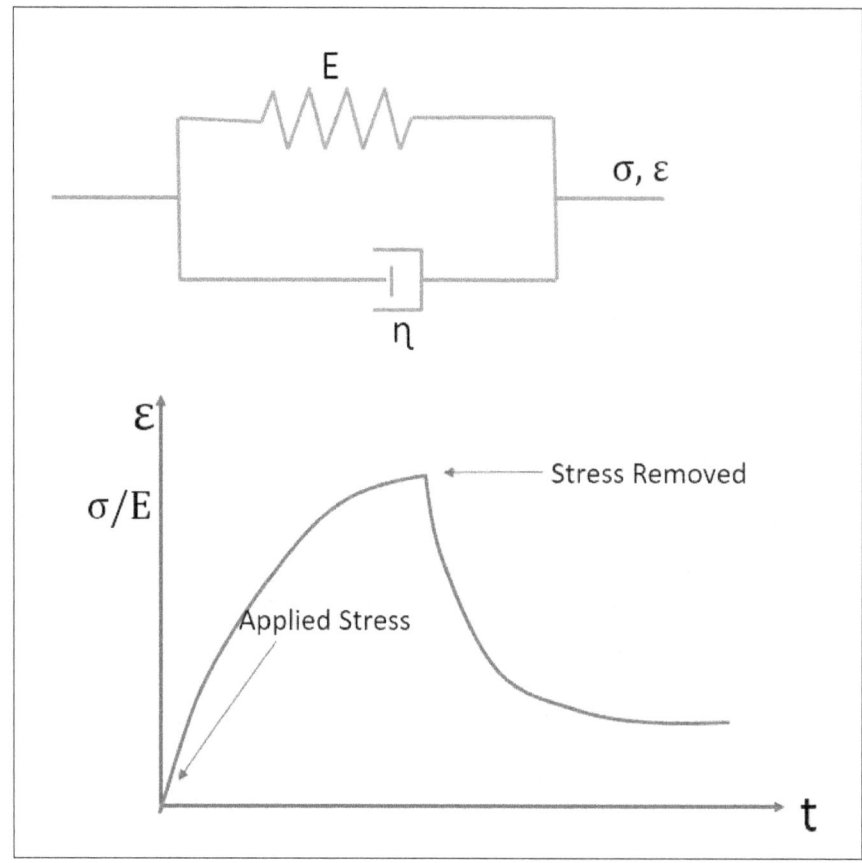

Figure .20: Kelvin-Voigt Model and its Creep Recovery Responsel

$$\sigma(t) = E\varepsilon(t); \quad (.15)$$

or

$$\frac{d\sigma}{dt} = E\frac{d\varepsilon(t)}{dt} \quad (.16)$$

The equation can be applied either to the shear stress or normal stress of a material. Newtons law for a viscous liquid states that;

$$\sigma(t) = \eta\frac{d\varepsilon(t)}{dt}; \quad (.17)$$

Combining equations (.15 and .17), we get the Kelvin-Voigt model for a linear viscoelastic material;

$$\sigma(t) = E\varepsilon(t) + \eta \frac{d\varepsilon(t)}{dt}; \qquad (.18)$$

There is a maximum deformation of the Kelvin-Voigt body for a given stress, provided that the system is loaded for a sufficiently long time. Thus making the KelvinVoigt model effective for predicting creep, but with limitations when describing the relaxation behavior of the material, once the stress load is removed permanently.

## .9.2 Maxwell Model

The description of a polymer material by ideal linearized elastic and viscous element can be successfully used to model the mechanical behavior of some real world materials. A simple approach to such description is the connection between these two linearized elements in series. The description of a material with this behavior is termed a Maxwell model.

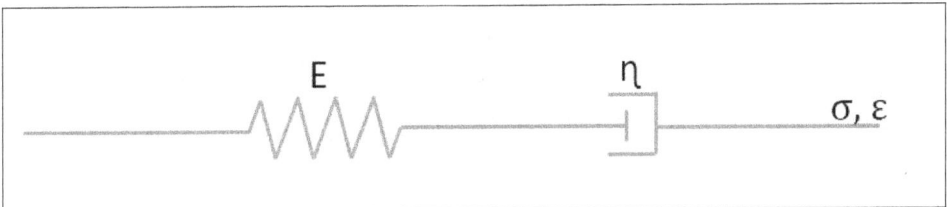

Figure .21: Maxwell Model

When a force is applied to a material, it goes through an elastic deformation immediately as per the Young's modulus. The deformation due to the viscous behavior of dashpot is initially zero and increases linearly with time under the constant applied load. After removal of the load an instantaneous recovery of elastic deformation takes place in the spring. The recovery in the dashpot is delayed and results in a permanent deformation even after removal of the load.

This permanent deformation exists because there is no restoration force providing the momentum to the viscous liquid in the dashpot to return to its original state. A Maxwell body is characterized by its inability to return to its original dimensions after the application and complete removal of loads. The total deformation at any time in the system is the sum of the two individual deformations, 1) The time-dependent viscous dashpot and 2) the time-independent spring.

$$\gamma(t) = \frac{\tau}{G} + \frac{\tau}{\eta}t = \tau\left(\frac{1}{G} + \frac{t}{\eta}\right); \qquad (.19)$$

Maxwell model shows that the resulting strain consists of two different, physically inde-

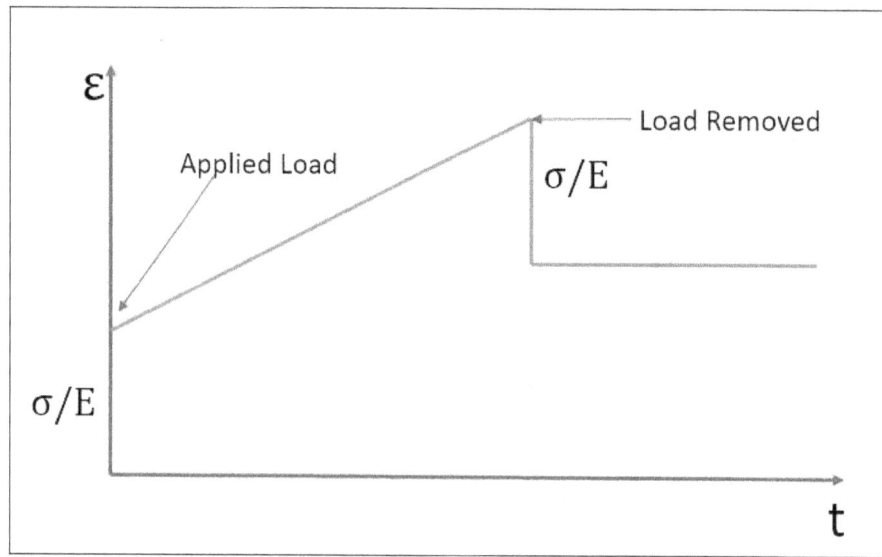

Figure .22: Creep Recovery Response in the Maxwell Model

pendent strains and there is no interaction between them.

## .9.3 Standard Linear Zener Model

The standard linear also known as the Zener model, is a method to model the behavior of a viscoelastic material. Kelvin-Voigt and Maxwell model prove insufficient when describing creep and stress relaxation. Zener model is one of the simplest models that predicts both phenomena. The standard linear Zener model combines properties of the Maxwell and KelvinVoigt models to accurately describe the overall behavior of a system under a given set of loading conditions. The behavior of a material applied to an instantaneous stress is shown as having an instantaneous component of the response.

Although Zener model can be used to accurately predict the general shape of the strain curve, as well as behavior for long term and instantaneous loads, the model lacks the ability

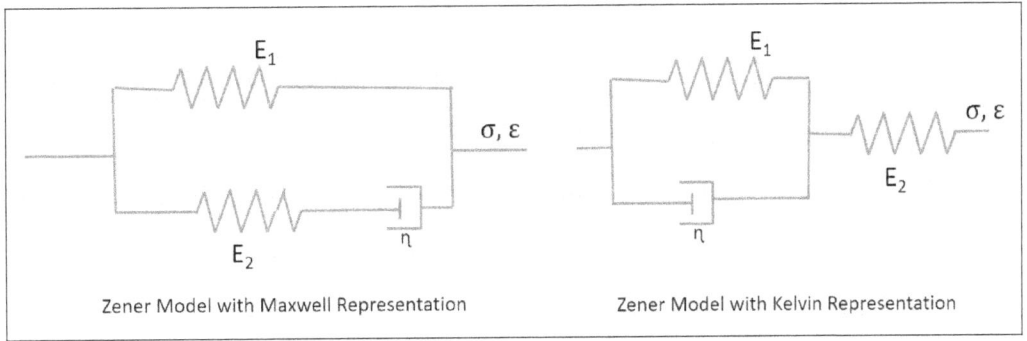

Figure .23: Zener Model with Maxwell and Kelvin Representations

to accurately model material systems numerically for application in computational mechanics.

## .10 Effect of Fillers on Dynamic Properties

Dynamic properties of rubbers are affected by fillers and other compound ingredients. It is found that the network can substantially increase the effective volume of the filler due to rubber trapped in the agglomerates, leading to a high elastic modulus. Fillers like carbon blacks increase the modulus of rubbers by forming reinforcing bonds with the polymers. The increase in modulus is due to the fact that fillers can be considered having higher rigidity as compared to the rubber. They do not participate in the deformation and increase the strain in the rubber matrix between adjacent filler particles. The strain increase because of this can be expressed by

$$X = 1 + 2.5c + 14.1c^2, \qquad (.20)$$

where $X$ is the strain increase ratio and $c$ is the volume fraction of the material. The equation shows that $tan\delta$ increases linearly with $x$.

## .11 Applications of Dynamic Material Characterization

Dynamic material characterization is a technique that measures stress as a function of strain, or force as a function of displacement and time. It also involves the application of one or more forces at various frequencies, as a means of determining how material changes in a dynamic environment where the material comes under the effect of multiple frequencies. Dynamic characterization testing of engineering components normally includes components like tires, springs, dampers, biomedical implants and vibration isolation components from the automotive, aerospace and biomedical industries. These components perform under time and frequency varying conditions during their entire lifecycle making them ideal candidates for dynamic study both for product development and failure analysis.

## .11.1 Development and Testing of Anti-Vibration Mounts

High vibration levels can cause machinery and component failures as well as objectionable noise levels. A common source of objectionable noise in buildings is the vibration of machines that are mounted on floors or walls. In automotive systems, undamped vibrations in suspension components and engine mounting systems can decrease ride comfort and cause pre-mature fatigue failures. A typical problem is a rotating machine such as a pump, AC compressor, blower, engine, etc mounted on a automotive chassis, on the floor or on the roof.

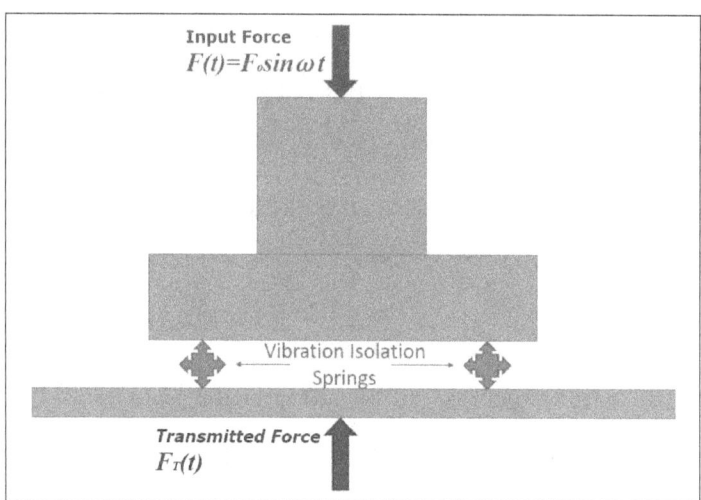

Figure .24: Transmissibility of Engine Mounts of Different Materials

If we consider only the vertical (axial) motion of an anti-vibration engine mount, the case shown in Figure(.24) can be described mathematically by a single degree of freedom, lumped element system.

$$M\ddot{x} + C\dot{x} + Kx = F(t) \quad (.21)$$

M = mass of system, K = stiffness, C = viscous damping, x(t) = vertical displacement, F(t) = excitation force.

If we neglect damping, the vertical motion of the system, x(t) can be shown to be;

$$W_n = \frac{K}{m} \quad (.22)$$

The system has a natural, or resonant frequency, at which it will exhibit a large amplitude of motion, for a small input force. In units of Hz (cycles per second), this frequency, $F_n$ is

$$F_n = \frac{1}{2\pi}\sqrt{\frac{k}{m}} \quad (.23)$$

Transmissibility for a vibration isolation mount is defined as the ratio of transmitted force to input force.

$$T_r = \frac{F_T}{F_I} \quad (.24)$$

At very low or idling frequencies the transmitted force is equal to the input force. As frequency increases, so does transmissibility, until a peak is reached at the resonant frequency (Natural frequency) of the system. At and around the resonance point, the mount is amplifying and transmitting a larger force than the input force. As frequency increases beyond the resonance point, the effectiveness of the mount increases and transmissibility is decreased leading to isolation of the vibration.

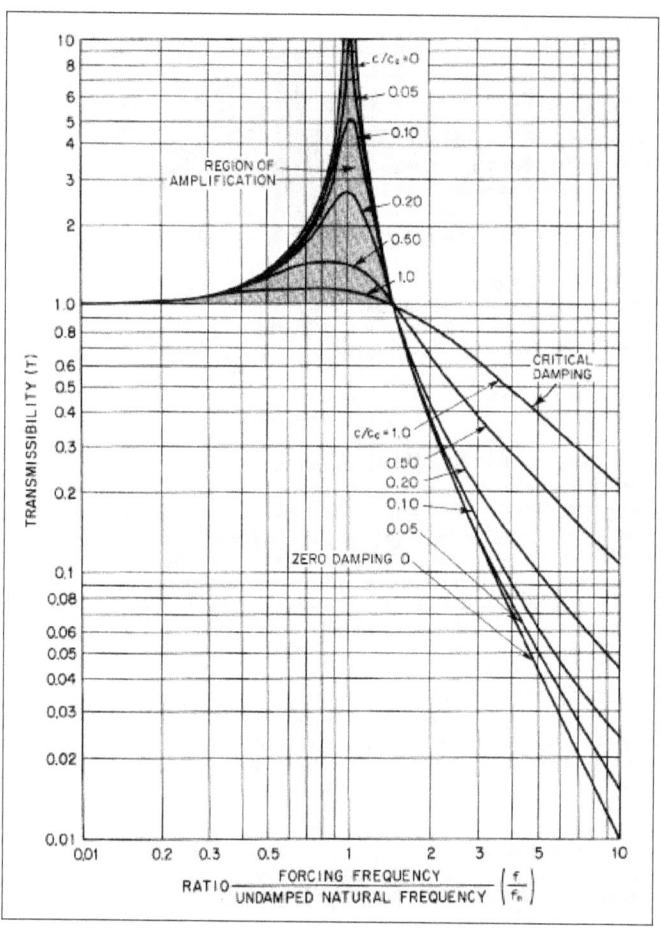

Figure .25: Transmissibility for a Viscously Damped Single Degree of Freedom System

Figure(.25) shows the transmissibility curve with region of amplification, area on the x-axis beyond ($\sqrt{2}$) frequency ration is the area of isolation. If the mass of the mount system is held constant, the resonant frequency of the system depends primarily on the dynamic stiffness of the mount. The softer the mount the lower the resonant frequency and any vibration beyond the resonant point is damped. Since the mount can perform as an isolator only at frequencies higher than resonance, mounts as soft as possible should be designed but with the caveat that softer materials may also lead to more hysteresis and heat generation leading to failures. The height of the peak at resonance is determined only by the damping in the mount and the material. The higher the damping, the lower the resonance peak, and the better the performance of the mount in the range of natural frequency. The effectiveness of a mount in reducing transmissibility at high frequencies is a function of the dynamic stiffness and damping. The response of elastomer mounts in this part of the frequency range is complex since elastic spring rate and damping can themselves be functions of frequency. Figure(.26) shows the transmissibility characteristics of a variety of commonly used elastomers. The effectiveness of the isolator, expressed in percent is;

$$\%Isolation = (1-T)*100 \tag{.25}$$

The transmissibility including the effect of damping is given by;

$$T_r = \frac{\sqrt{1+\zeta r^2}}{\sqrt{(1-r^2)^2 + (2\zeta r)^2}} \tag{.26}$$

where $\zeta = \frac{c}{2m\omega_n}$ = critical damping ratio.

All the compounds tested were formulated to 55 +-5 Shore A Hardness. All the mounts were tested under the same conditions of frequency sweep with a pre-load on the sample. As shown the resonant frequencies of all the mounts tested fell within a fairly narrow range.

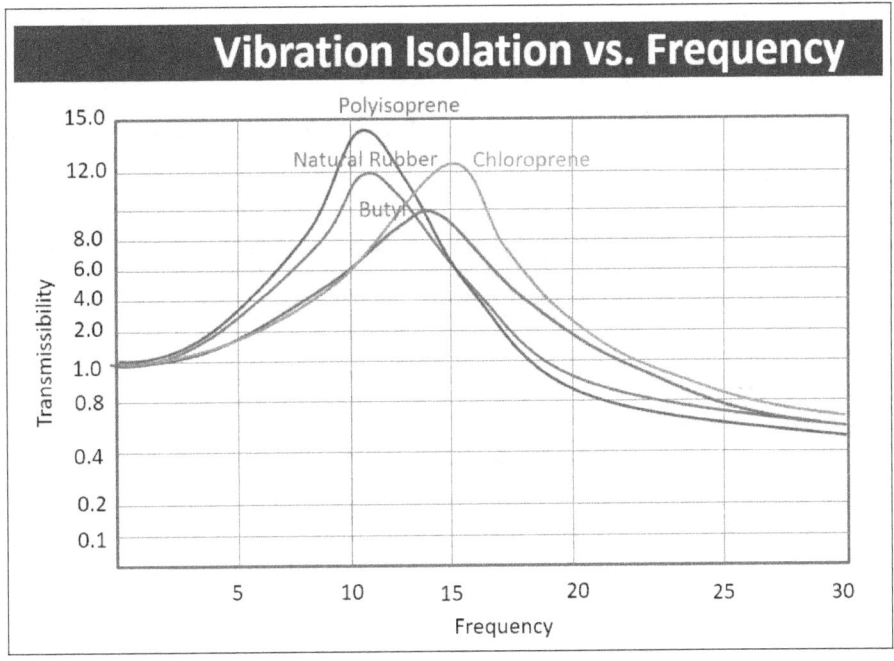

Figure .26: Transmissibility of Engine Mounts of Different Materials

Natural rubber (NR) is one of the most widely used material for vibration isolation applications. This elastomer is the benchmark for comparison of most other elastomers. It was the first elastomer and has some uniquely desirable properties in anti-vibration applications. Natural rubber has high strength, when compared to most synthetic elastomers. It has excellent fatigue properties and low to medium damping which translates into efficient vibration isolation. Natural rubber is not very sensitive to vibration amplitude levels but it is susceptible to weathering from ozone. At the high temperature end, natural rubber is often restricted to use below approximately 120°C.

## .11.2 Material Characterization Testing under High Strain Rates

At our laboratory we have studied the mechanical behaviour of High Density PolyEthylene(HDPE) polymer under the effect of various temperatures and strain rates. Mechanical characterizations are carried out through uniaxial compression and split Hopkinson pressure bar (SHPB) for revealing low and high strain rate response, respectively. Meanwhile, the experiments are performed for strain rates varying from and a temperature range of 313°K to 393° K. The experimental results reveal that the stress-strain behaviour of HDPEs is much different at lower and higher strain rates. At higher strain rate, the HDPEs yield at higher stress compared to that at low strain rate. At lower strain rate, yield stress increases with the increase in strain rate while it decreases significantly with the increase in temperature. Likewise, initial elastic modulus increases with the increase in strain rate. Yield stress increases significantly at higher strain rates in the material.

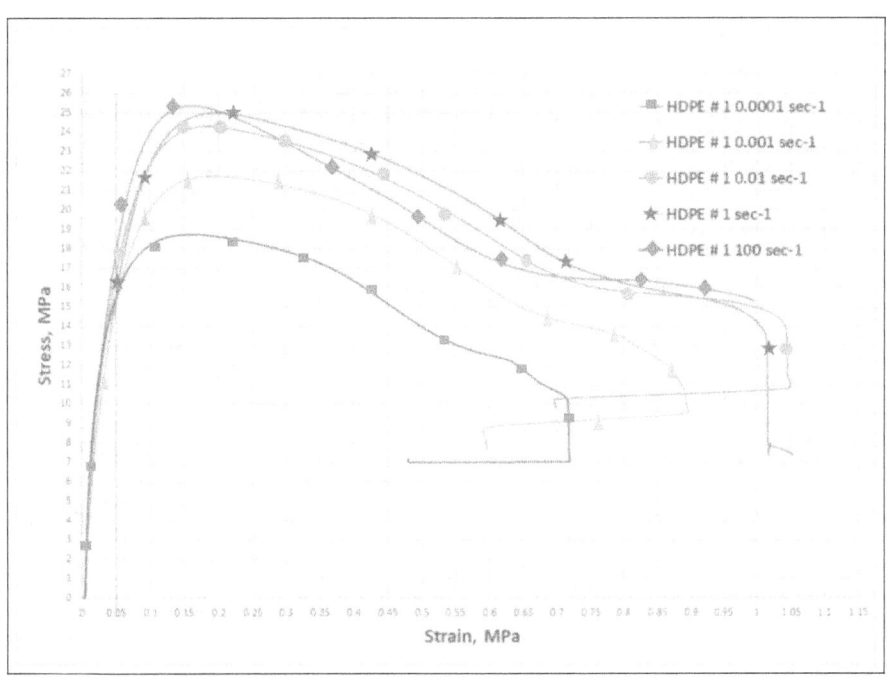

Figure .27: High Density PolyEthylene(HDPE) polymer under the effect of different strain rates

Uniaxial tensile tests were performed to determine the dynamic response of HDPEs at strain rates varying from 0.0001 $sec^{-1}$ to 100 $sec^{-1}$. Dynamic tests were performed at seven different strain rates, and the results in terms of true stress-strain curves are shown in Figure .27. The results show that yield stress increases with the increase in strain rate. The stress-strain curves show almost similar mechanical response in which initial nonlinear elastic behaviour was observed followed by subsequent yielding, strain softening and hardening. Yield stress changes significantly with the increase in strain rate. An increase of 20.6 % in yield stress was calculated with strain rate increase from 0.0001 $sec^{-1}$ to 100 $sec^{-1}$. At all strain rates, ductile response of HDPEs was observed. Strain-rate dependency of the stress-strain behaviour of polymer materials has now been well documented. This feature of mechanical behaviour is important in engineering applications for automotive and aerospace crashworthiness where the design of a polymer component is required to resist shock and impact loading and other strength stiffening effects.

## .11.3 Development and Failure Analysis of Rubber Rollers

Equation (.20) can be particularly applied to materials where the carbon black filler size is larger (e.g. N990). In smaller sized filler particles the rubber is just in static form around the blacks and effective volume fraction has to be suitably used. The strain increase or amplification concept due to the presence of fillers also affects energy losses per cycle of operation. Since the presence of higher rigidity fillers magnify and increase the local strain, the dynamic losses which are proportional to local strain amplitudes also increase. During cyclic strain, while the stable filler network can reduce the hysteresis of the filled rubber, the breakdown and reformation of the high structure filler network would cause an additional energy dissipation resulting in higher hysteresis. More information on this increase of losses can be obtained from *Meinecke et al.* As per studies, as compared to carbon black, silica is able to form a stronger and more developed filler network resulting in higher modulus and lower hysteresis at low and room temperature applications.

The E', E", E* or Tan Delta values are to be used as a comparative set of values from different compounds, or a single compound tested at different conditions i.e., temperature, frequency, or strain levels. When different compounds are tested, variables such as cure systems, filler types and levels and plasticizers can be evaluated and compared to provide inputs about dynamic properties.

For an example where dynamic properties testing could be helpful a filler is changed from N-762 to a N-550 black in a rubber covered pinch roller for paper applications. Some other compound adjustments were also made to maintain the same tensile, elongation and durometer values. A few weeks after the compound revision it was observed that there is a noticeable increase in failures of the roller. The operating conditions dictate that the paper roller runs at 180 rpm under 8000 Newtons resulting in the rubber warming up. A dynamic test at three Hertz and Tan Delta values is carried out to compare the compounds. The greater values of E" and Tan delta (damping coefficient) indicate a higher hysteresis in the compound. It can be inferred from the results that higher hysteresis can cause greater heat buildup in the rollers leading to failures.

## .11.4 Viscoelastic Analysis of Tires

Tires are subjected to high cyclical deformations when vehicles are running on the road. When exposed to harsh road conditions, the service lifetime of the tires is jeopardized by many factors, such as the wear of the tread, the heat generated by friction, rubber aging, and others. As a result, tires usually have composite layer structures made of carbon-filled rubber, nylon cords, and steel wires, etc. In particular, the composition of rubber at different layers of the tire architecture is optimized to provide different functional properties. The desired functionality of the different tire layers is achieved by the strategical design of specific viscoelastic properties in the different layers. Zones of high loss modulus material will absorb energy differently than zones of low loss modulus. The development of tires utilizing dynamic characterization allows one to develop tires for smoother and safer rides in different weather conditions.

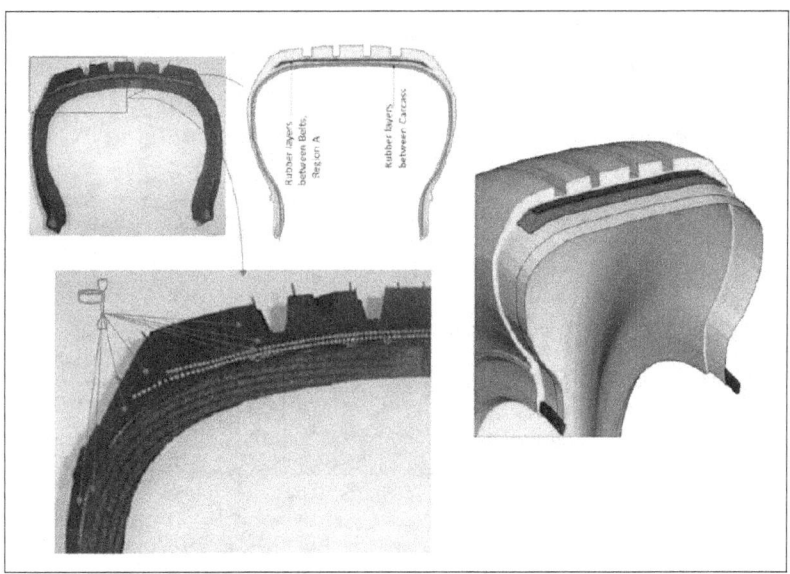

Figure .28: Locations of Different Materials in a Tire Design

The dynamic properties are also related to tire performance like rolling resistance, wet traction, dry traction, winter performance and wear. Evaluation of viscoelastic properties of different layers of the tire by DMA tests is necessary and essential to predict the dynamic performance. The complex modulus and mechanical behavior of the tire are mapped across the cross section of the tire comprising of the different materials. A DMA frequency sweep test is performed on the tire sample to investigate the effect of the cyclic stress/strain frequency on the complex modulus and dynamic modulus of the tire, which represents the viscoelastic properties of the tire rotating at different speeds. Significant work on effects of dynamic properties on tire performance has been carried out by Ed Terrill et al. at Akron Rubber Development Laboratory, Inc.

## .11.5  Non-linear Viscoelastic Tire Simulation Using FEA

Non-linear Viscoelastic tire simulation is carried out using Abaqus to predict the hysteresis losses, temperature distribution and rolling resistance of a tire. The simulation includes several steps like (a) FE tire model generation, (b) Material parameter identification, (c) Material modeling and (d) Tire Rolling Simulation. The energy dissipation and rolling resistance are evaluated by using dynamic mechanical properties like storage and loss modulus, tan delta etc. The heat dissipation energy is calculated by taking the product of elastic strain energy and the loss tangent of materials. Computation of tire rolling is further carried out. The total energy loss per one tire revolution is calculated by;

$$\Psi^{diss} = \sum_{i=1}^{\infty} i 2\pi \Psi_i Tan\delta_i, \qquad (.27)$$

where $\Psi$ is the elastic strain energy,
$\Psi^{diss}$ is the dissipated energy in one full rotation of the tire, and
$Tan\delta_i$, is the damping coefficient.

The temperature prediction in a rolling tire shown in Fig (.29) is calculated from the loss modulus and the strain in the element at that location. With the change in the deformation pattern, the strains are also modified in the algorithm to predict change in the temperature distribution in the different tire regions.

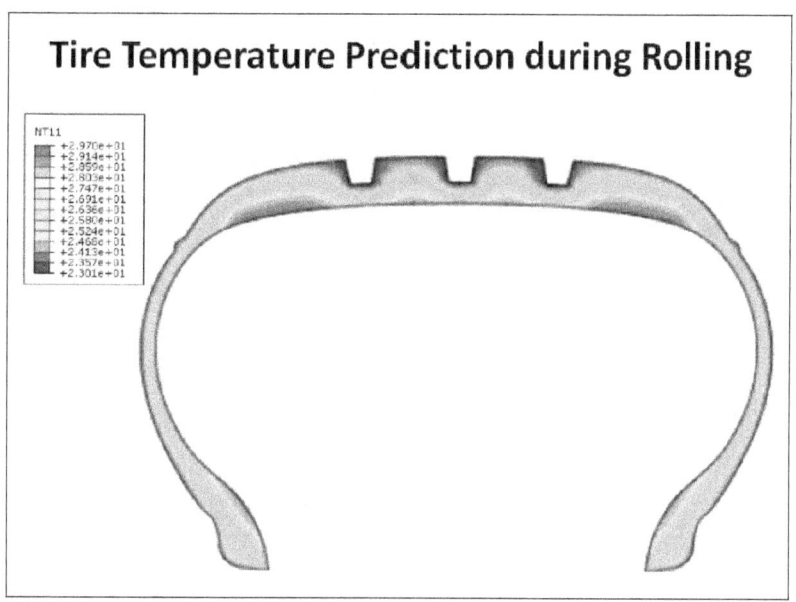

Figure .29: Temperature Distribution in a Tire under Rolling Conditions

Rolling resistance is now calculated from;

$$F_{RR} = \frac{-\Psi^{diss}}{2\pi R} where,$$ (.28)

$F_{RR}$ is the Rolling Resistance

$2\pi R$ is the Circumferential length.

In summary, the absolute values from dynamic tests are meaningful, but have little utilitys as isolated data points. They do become valuable data points when compared to each other or some other known variable. A tan delta or damping coefficient value of 0.4 is poor for a natural rubber or EPDM based compound, but very good in FKM materials where the structure of the compound makes it venerable to lower than optimum dynamic properties. Most uncured rubbery compounds start on the viscous side, and as we cure the compound, we shift towards the elastic side.

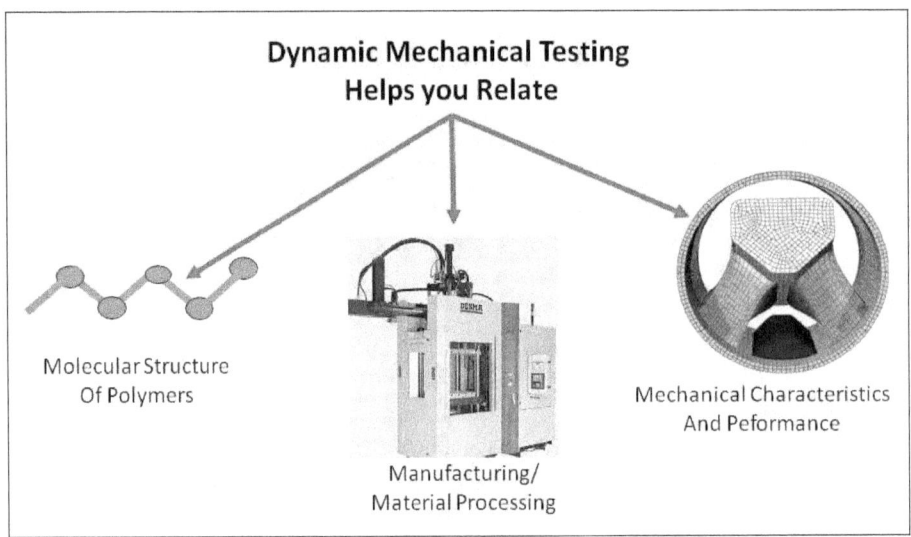

Figure .30: Viscoelastic Studies Correlate Molecular Structure to Manufacturing and Mechanical Properties of Engineering Components

As it stands today, the theory of dynamic properties can be applied judiciously to product development or failure analysis problems. The field of application has evolved over time with availability of highly sophisticated instruments. The problems need to be studied upfront for any time or frequency dependent loads and boundary conditions acting on the components and the theory be suitably applied. Needless to say that dynamic properties have utmost importance when rubber components show heat generation and fatigue related field failures as it relates the molecular structure of the polymer material to the manufacturing process and to the field performance of engineering products.

## .12 The Correlation Between Frequency and Temperature

A similarity in the mechanical properties of polymers is noticeable both at low temperatures and at high frequencies. This similarity is due to the viscoelastic nature of polymer materials. If a force is applied on an elastomer for varying periods of short and long time duration, there is a correlation between the frequency of the applied load and the temperature at which it is applied. Because of the high viscous nature of elastomers, the stress that occurs during deformation is partially reduced by chain displacements and entropy absorption. The time required for this is known as relaxation. As the temperature decreases, the chain movements in the polymer slowly decrease and the length of time needed to decrease the stress increases. The elastomer slowly hardens and becomes glassy. If the load frequency increases, the chain mobility is no longer sufficient for absorption of the deformation. The material appears to be glass like solid. This is known as dynamic hardening. This means that an increase in frequency has the exact same effect as a drop in temperature due to which the material hardens. Accordingly, the glass transition temperature $T_g$ depends not only on the method of determination, but also on the test frequency and applied temperature. This effect is known as the time-temperature correspondence. The result of this phenomenon is that an increase in frequency at low temperatures leads to a premature hardening of the material. This influence of relaxation, which depends on temperature and frequency, must therefore be taken into account when designing engineering components.

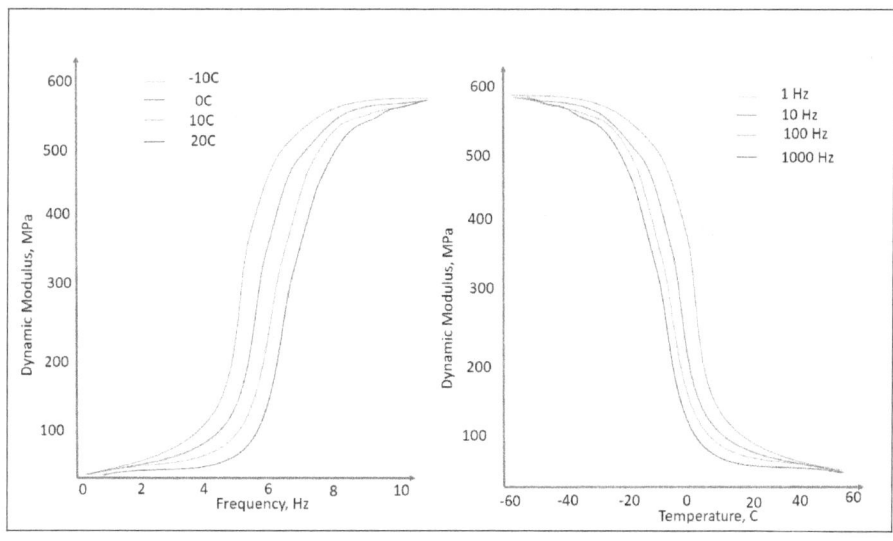

Figure .31: Dependency of Dynamic Modulus on Frequency and Temperature

## .13 Time-Temperature Superposition (TTS)

TTS is a valuable tool for describing the viscoelastic behaviour of linear polymers over a large range of frequencies and time. Viscoelastic properties like G', G", modulus, creep and relaxation compliance etc., are the experimental parameters. TTS involves the use of a temperature dependent shift factor for the magnitudes of measured stress (vertical shift factor) and for time or frequency (horizontal shift factor). The aim of TTS is to generate a "Master Curve" showing viscoelastic properties over much larger range of times and frequencies that can be studied using the limited experimental resources.

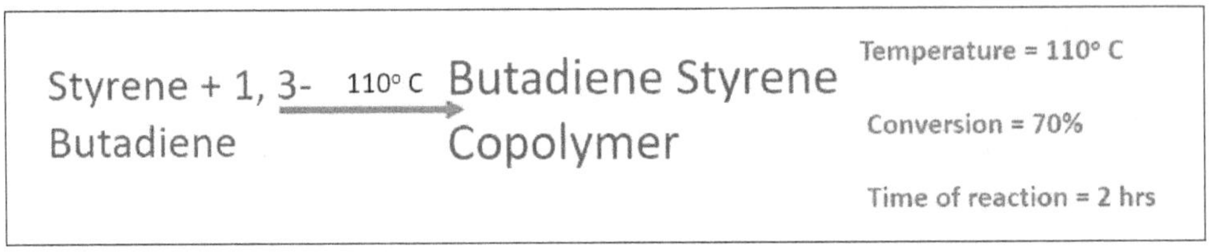

Styrene + 1, 3- Butadiene $\xrightarrow{110°C}$ Butadiene Styrene Copolymer

Temperature = 110° C
Conversion = 70%
Time of reaction = 2 hrs

Figure .32: Reaction Equation for Styrene Butadiene

As shown in in Figure(.32) during a polymer reaction if the desired conversion is 80%, then this conversion can be obtained either by increasing the temperature or by increasing the time of reaction. Polymer properties obtained at a particular combination of time and frequency and temperature, can also be obtained at some other combinations of time frequency and temperature. The technique used to calculate and predict is known as time-temperature superposition.

With viscoelastic materials, time and temperature are equivalent to the extent that data at one temperature can be superimposed on data at another temperature by shifting the curves along the log time axis.

Following are the main features of the TTS principle;

1. Enables evaluating long time behaviour by measuring stress relaxation or creep data over a shorter period of time but at several different temperatures.

2. TTS is an *empirical* relationship.

3. TTS is based on the observation that, for a single material, the curves of the viscoelastic properties, generated at different temperatures, are similar in shape when plotted against log time or log frequency. The Curves generated at different temperatures can be exactly superimposed by shifting along these axes.

4. TTS is applicable purely because of the viscoelastic nature of polymers and applies to stress relaxation, creep and dynamic mechanical measurements.

5. Information from each temperature curve is combined to yield a master curve at a single temperature.

Following are the main uses of the Time-temperature superposition (TTS) principle;

1. To determine the temperature dependence of the rheological behaviour of a polymer liquid.

2. To expand the time or frequency at a temperature at which the material is being characterised.

3. Any instrument, due to its mechanical limitations can usually provide data over a limited range of times or frequencies at a particular temperature.

4. Using low frequency (cheap) rheometer to predict high frequency (high) rheometer properties.

5. Cost of a MDR is US$10K, while Capillary rheomter is more than US$50K.

Low temperature corresponds to a high frequency response and vice-versa

**Assumptions in the Use and Application of TTS priniple are;**

1. The material does not undergo any chemical or physical changes as a result of the temperature change.

2. There is no phase change in the material as a result of change in temperature.

3. Applicable in linear viscoelastic regime only.

4. Material is categorized as "Thermorheologically Simple", can be simply said to be when y-axis and x-axis properties of TTS materials have the same temperature dependence.

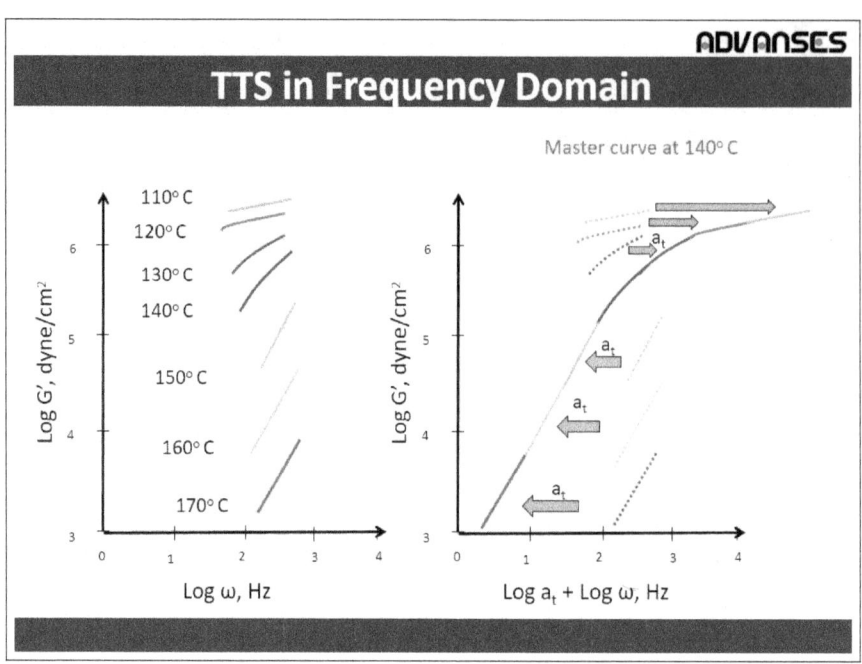

Figure .33: Time Temperature Superposition Shifting of Results to Desired Temperature

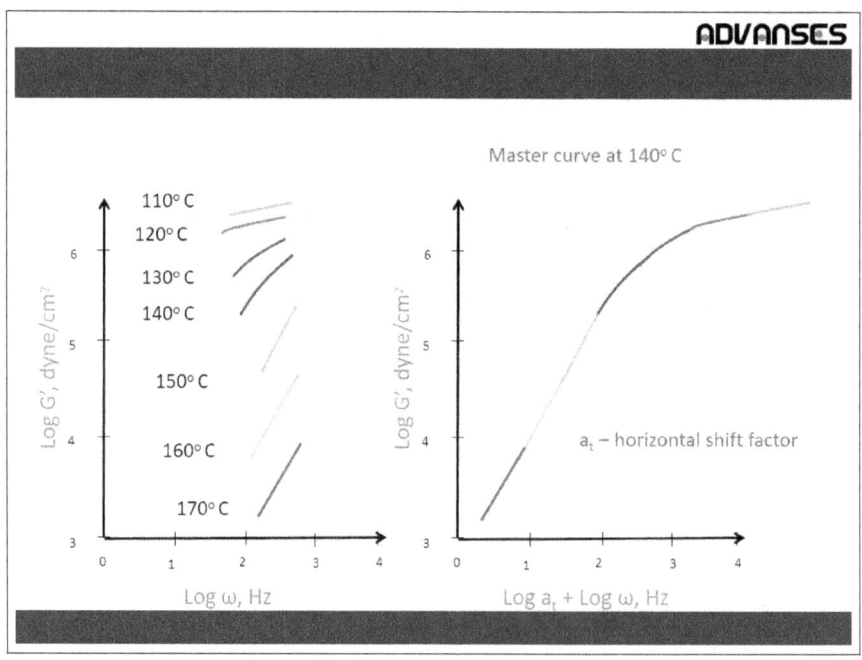

Figure .34: Developed Master Curve from TTS in the Frequency Domain

As shown in Figures(.33 and .34) the time temperature superposition technique is used to make a master curve that describes the property of a material at a specific temperature.

The benefit of using TTS as explained earlier is to use to generate the material properties for a big range of dynamic frequencies at different temperatures. It can be observed in Figures(.33 and .34) that approximately 7 tests have been carried out at temperatures of 112°C to 170°C, all these results at a smaller frequency range are then converted to a master at one 40°C by shifting the graphs from the left to the right using the shift factor $a_t$.

### .13.1 Calculation of shift factor

The WLF equation is typically used to describe the time-temperature behavior of polymers in the glass transition region. The equation is based on the assumption that, above the glass transition temperature, the fractional free volume increases linearly with respect to temperature. The model also assumes that as the free volume of the material increases, its viscosity rapidly decreases. The other model which is commonly used to relate the shift factors with respect to temperature is the Arrhenius relation:

**WLF equation**

$$Loga_t = \frac{-C1(T-To)}{[C2+(T-To)]} where, \qquad (.29)$$

C1 and C2 are constants

T is temperature in Kelvin

To is reference temperature in Kelvin

The values of C1=17.4 and C2=51.6K were originally thought to be universal and are still widely used.

**Arrhenius equation**

$$a_t = exp[(E_a/R)(1/T1/To)] where, \qquad (.30)$$

$E_a$ is activation energy

R is universal gas constant

The log Shift Factor ($a_t$) versus temperature plot should be a smooth monotonic curve as shown in Figure, provided the mechanism of relaxation remains the same during the process.

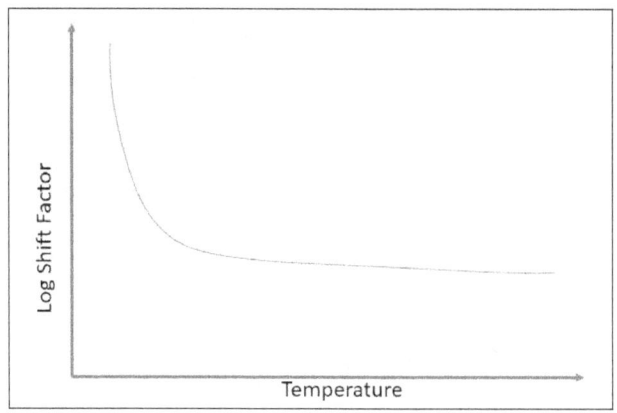

Figure .35: Shift Factor Verification

## .13.2 Generalized Steps for TTS

1. Pick a reference temperature

2. Multiply the time at each temperature by the constant that gives the best overlap with the reference temperature data

3. Define that multiple as $a_T$ ($a_T = 1$ for ref. temp.)

4. Find $a_T$ for each temperature

Plot $\log(a_T)$ vs $1/T$ linear if Arrhenius;

$$k = Ae^{\frac{-Ea}{RT}} = Empirical equation = ln(k) = ln(A)E_a/RT \tag{.31}$$

DMA Master Curves can be generated in any of the three modes - fixed frequency, stress relaxation, and creep.

Other examples of TTS use in engineering applications include:

1. Gaskets - to measure flow (creep) and stress relaxation effects which reduce seal integrity over time.

2. Force-fit snap parts - to measure stress relaxation effects, which can lead to joint failure.

3. Structural beams - to measure modulus drop with time, which leads to increasing beam deflection under load over time.

4. Bolted plates - to measure creep of the polymer, which reduces the stress applies by the fastener.

5. Hoses - to measure creep of the polymer, which can lead to premature rupture of the hose.

6. Acoustics - to aid in the selection of materials which exhibit high damping properties in a specific frequency range.

7. Elastomer mounts - to assess the long-term creep resistance of mounts used for vibration damping with engines, missiles and other heavy equipment.

TTS procedure is used to generate material characterization data over a broad range of frequencies and temperatures by shifting data points. This procedure results in a dramatic increase in the range of the time scale. The shift factors can be calculated using the WLF equations and Arrhenius equations. Softwares are available in the market to carry out the whole procedure including iterative curve fitting.

## .14 Instruments for Dynamic Testing of Polymers

There are five (5) main classes of experiments for measurement of viscoelastic behaviour

1. Transient measurements: creep and stress relaxation

2. Low frequency vibrations: free oscillations methods

3. High frequency vibrations: resonance methods

4. Forced vibration non-resonance methods

5. Wave propagation methods

The frequency scale for the different test methods are shown in Figure(.36)

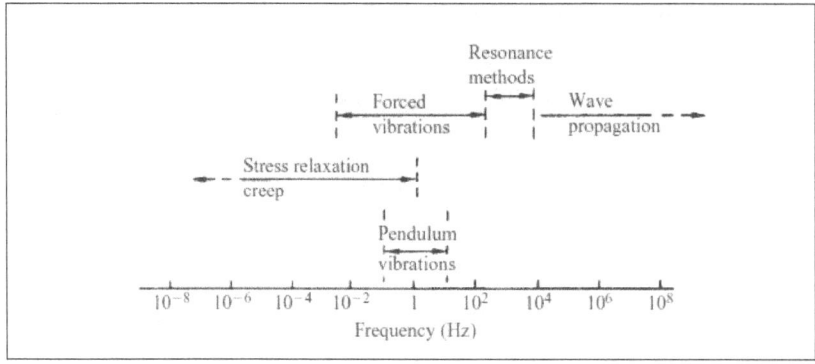

Figure .36: Frequency Scale for Different Test Methods

Dynamic characterization of elastomeric materials and polymers requires the use of sophisticated instruments with high fidelity load cells, displacement transducers and strain gauges to understand the deformations taking place in the material under dynamic frequencies. A Rheometer is used to test the dynamic properties during cure. A servo hydraulic fatigue testing machine and dynamic mechanical analyzer (DMA) instruments are the primary material testing instruments used in dynamic characterization of polymers. The sophistication of material testing instruments increase with needs for higher frequencies and higher loads. Strain gauge based and piezoelectric quartz based load cells are used in this instruments to study the load, stress and strain and record the test data.

Along with high quality hardware need also arises for a sophisticated and advanced software to carry out all the calculations. The software that calculates all the dynamic properties also needs to be as sophisticated and advanced as the hardware required to do the test

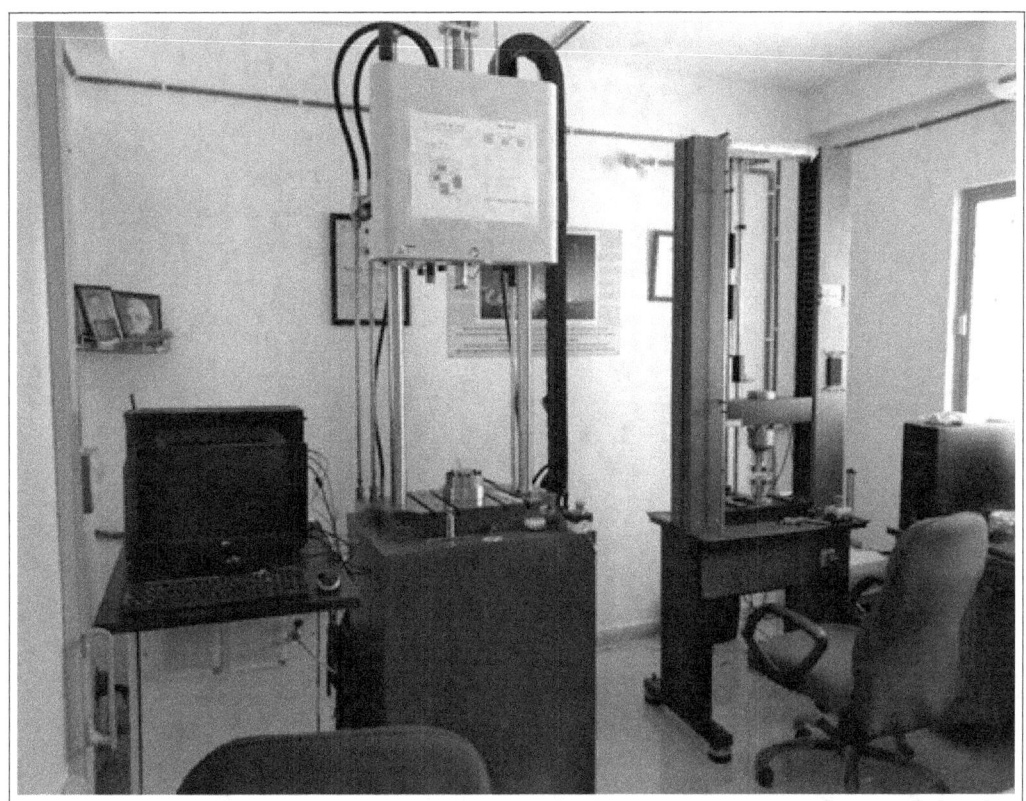

Figure .37: AdvanSES High Frequency and ElectroMechanical Testing Setup

Figure(.37) shows the AdvanSES servo hydraulic tester used in material testing at high frequencies the servo hydraulic tester is capable of going up to 100 hertz under sine wave definition hydraulic actuator is the primary source of frequency generation in the instrument and the servo valve in the actuator controls the flow of hydraulic fluid into the actuator so as to apply a controlled displacemtn at controlled frequency. The load cell in the instrument measures the loads generated in the sample under the dynamic frequencies. The servo hydraulic tester is primary primarily used to study static and dynamic stiffness, loss and storage modulus and Tan-delta. Fatigue crack growth propagation of rubber samples can also be tested using a high fps camera integrated with the tester. Elevated temperature testing is also available with the use of a temperature chamber with automatic PID control.

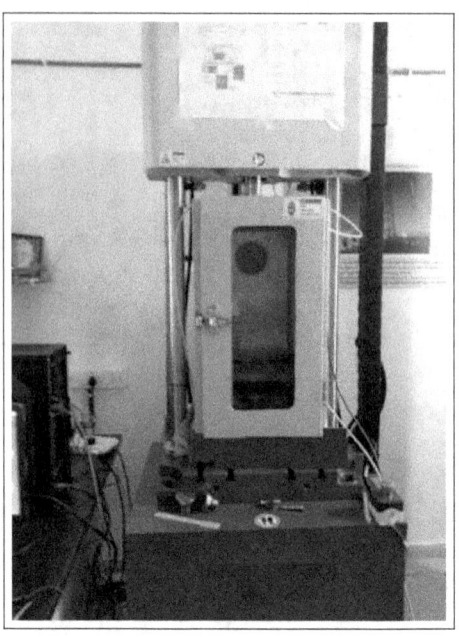

Figure .38: AdvanSES High Frequency Testing Setup with Temperature Chamber

High temperature tests (e.g. tensile, compression, flexure and fatigue tests) are used to determine the thermal-elastic behavior, heat resistance, endurance and durability of metallic and polymer materials. Elevated temperatures is combined with mechanical testing, environmental aging, analytical solution methods to develop and provide a comprehensive test protocol to evaluate materials and components.

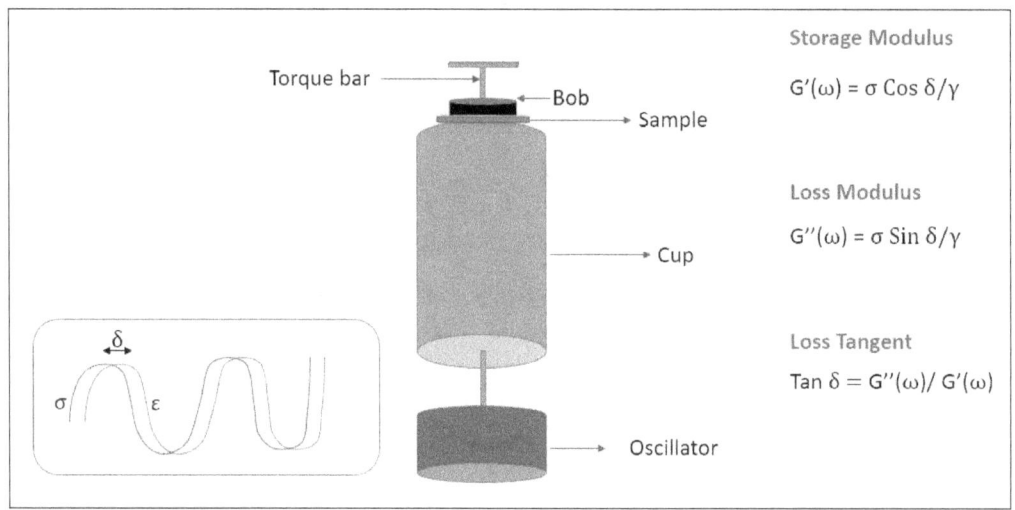

Figure .39: Oscillating Rheometer for Dynamic Testing

Figure(.39) shows a rheometer instrument used for studying the dynamic properties of uncured and cured rubber compounds. The instrument consists of a torque applicator, an oscillator and a load measuring device. The sample is held between the torque bar and the cup. The oscillator supplies an oscillatory motion that is transferred to the sample by the cup. The angular torsional deformations of the sample and the load generated in the instrument are measured using advanced load cell, displacement transducers and rotary encoders. A sine wave is input into the sample and a sine wave is similarly output from the instrument, both the input and output waves are compared to calculate the storage modulus, loss modulus and the tangent delta.

The sample requirements for the rheometer testing are that the samples should be dimensionally stable of rectangular or cylindrical cross section. To test the material a sample is firmly gripped at both ends. the specimen is electromagnetically or servo driven into sinusoidal oscillations of defined amplitude and frequency. The viscoelastic properties of the material makes the torque lag behind the deformation. The lag between the input and output is the phase angle as shown earlier in Figure(.12). The observed values for load, phase angle, and geometry constant of the specimen is used to calculate the complex shear modulus $G^*$, the storage shear modulus $G'$, the loss shear modulus $G''$, and tan delta.

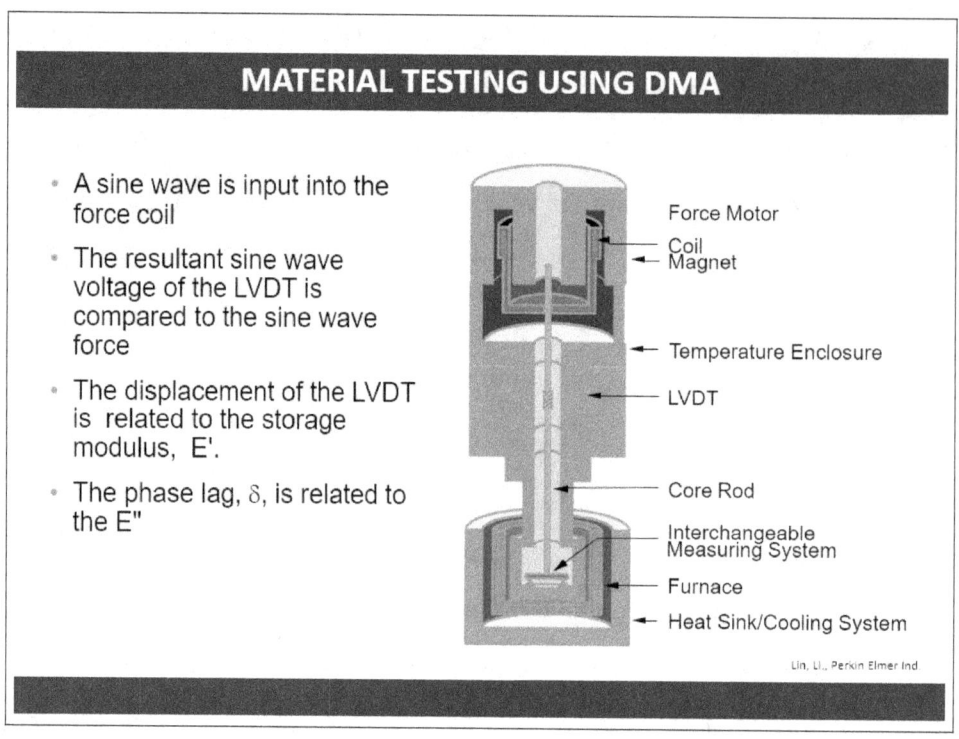

Figure .40: Axial Dynamic Mechanical Analyzer with Furnace

Figure(.40) shows a DMA instrument manufactured buy TA Instruments. As shown in the figure, a force motor with a coil and magnet is used to apply a force and lvdt measures the displacement of the sample. Furnace is provided for elevated and low temperature measurements. The sample is kept inside the furnace and a sine wave is input into the force motor, the resultant sine wave voltage of the lvdt is now compared to the input sine wave and the storage modulus, loss modulus, phase and tan delta are calculated from the test results data.

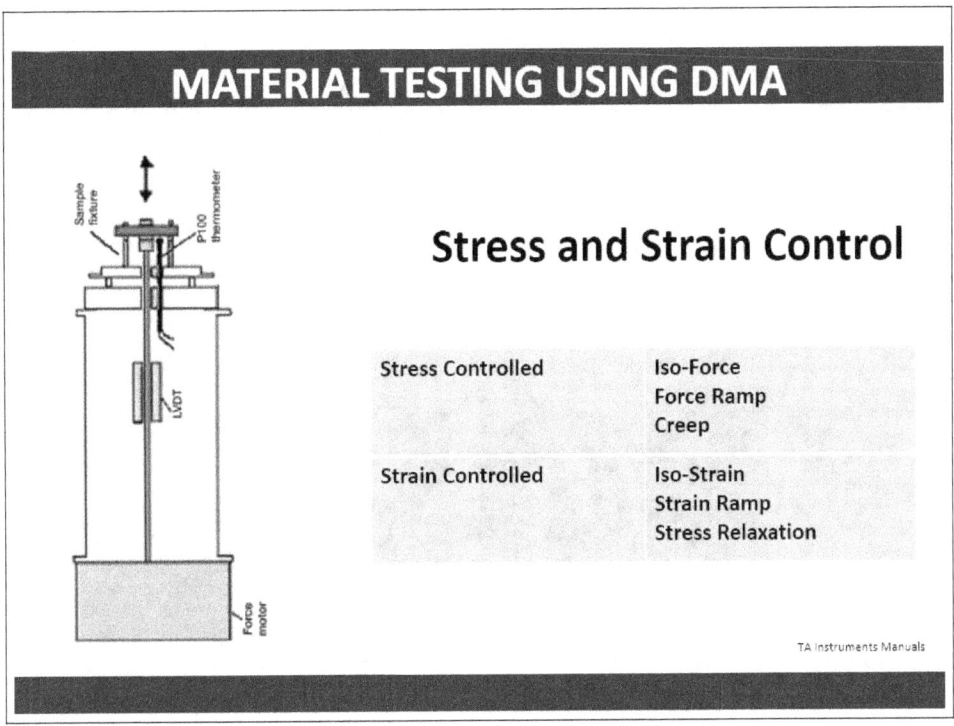

Figure .41: Axial Dynamic Mechanical Analyzer with Interchangable Sample Fixtures. Image Courtesy: Perkin Elmer Industries

Figure(.41) shows a DMA instrument from Perkin Elmer. When compared to the TA Instruments, both the machines have similar performance. Both the instruments can be operated under stress control and strain control. Creep and stress relaxation experiments can be carried out in all the instruments along with frequency sweep, strain sweep and temperature sweep studies

A DMA instrument is very versatile instrument able to apply different deformation modes on the sample. Different deformation modes can be chosen based upon the quality of the material and the material properties under study. Figure(.42) shows the different deformation modes available for application in a DMA instrument. Single and double cantilever beam, tensile, shear, compression and three point bending tests can be carried out using different kinds of fixtures.

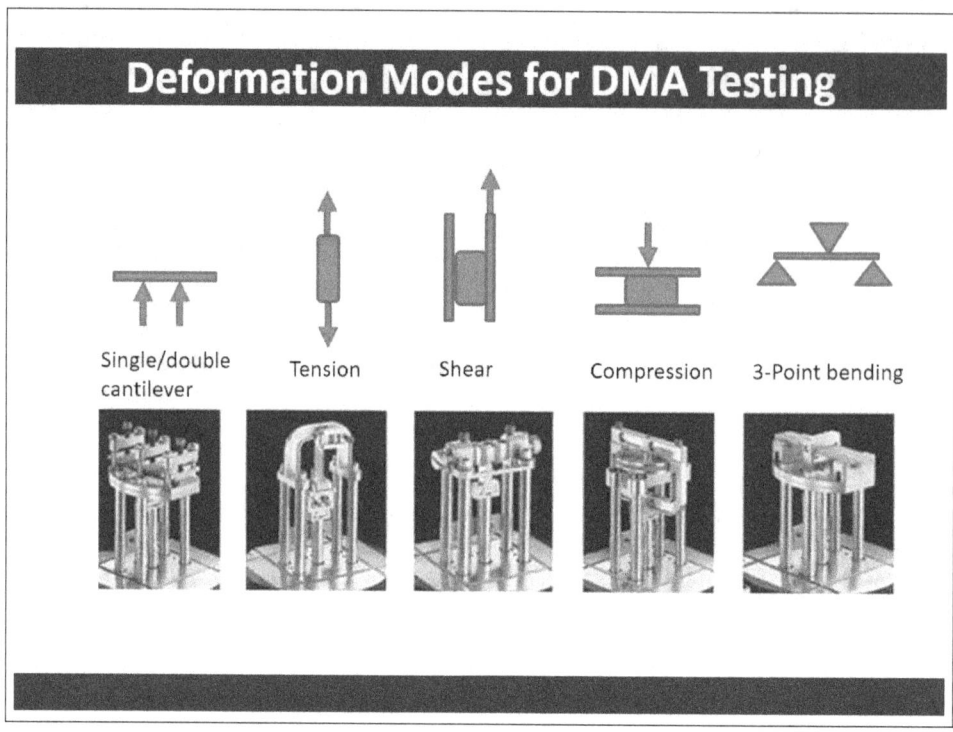

Figure .42: Deformation Modes Available in a Typical DMA Machine, Image Courtesy: TA Instruments and Perkin Elmer Industries

Materials such as hard polymers or soft viscoelastic elastomers are ideal materials to be tested on a DMA machine for dynamic properties. The testing conditions and parameters such as applied frequency range, temperature and available sample sizes and shape dictate the machine required for the testing. To carry out frequency and strain sweep studies on automotive and aerospace components, it becomes imperative to use a servo hydraulic machine. While the necessity to study the dynamic property of a material during processing makes it imperative to use a moving or oscillating die rheometer.

The importance of dynamic testing comes from the fact that performance of elastomers and elastomeric products such as engine mounts, suspension bumpers, tire materials etc., cannot be fully predicted by using only traditional methods of static testing. Elastomer tests like hardness, tensile, compression-set, low temperature brittleness, tear resistance tests, ozone resistance etc., are all essentially quality control tests and do not help us understand the performance or the durability of the material under field service conditions. An elastomer is used in all major applications as a dynamic part being able to provide vibration isolation, sealing, shock resistance, and necessary damping because of its viscoelastic nature. Dynamic testing truly helps us to understand and predict these properties both at the material and component level.

## .15 ASTM D5992 and ISO 4664-1

ASTM D5992 covers the methods and process available for determining the dynamic properties of vulcanized natural rubber and synthetic rubber compounds and components. The standard covers the sample shape and size requirements, the test methods, and the procedures to generate the test results data and carry out further subsequent analysis. The methods described are primarily useful over the range of temperatures from cryogenic to 200°C and for frequencies from 0.01 to 100 Hz, as not all instruments and methods will accommodate the entire ranges possible for material behavior.

Figures(.43 and .44) show the results from a frequency sweep test on five (5) different elastomer compounds. Results of Storage modulus and Tan delta are plotted.

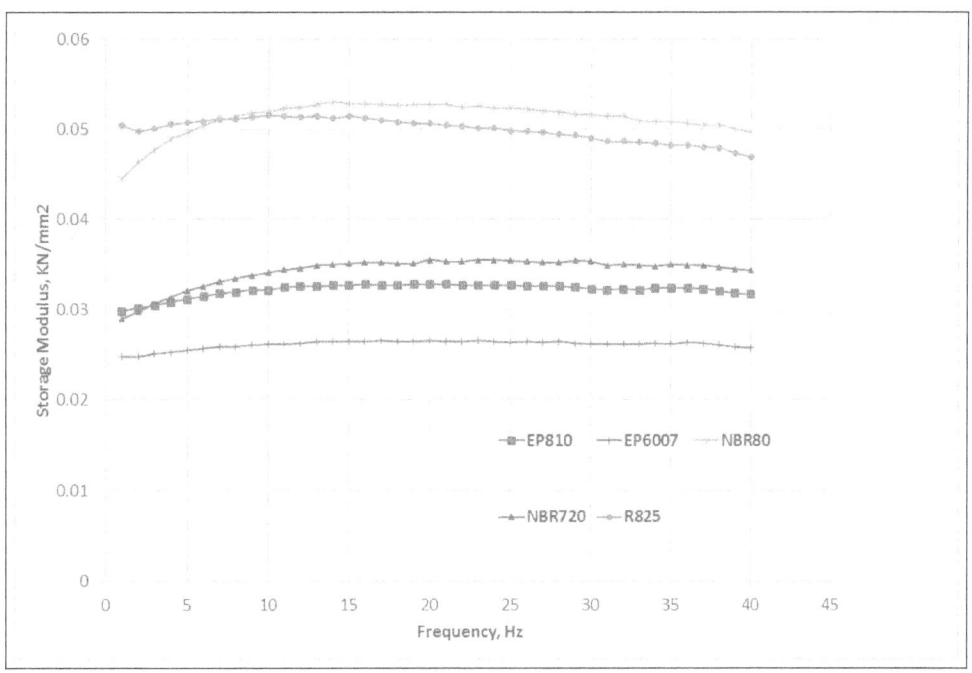

Figure .43: Plot of Storage Modulus Vs Frequency from a Frequency Sweep Test

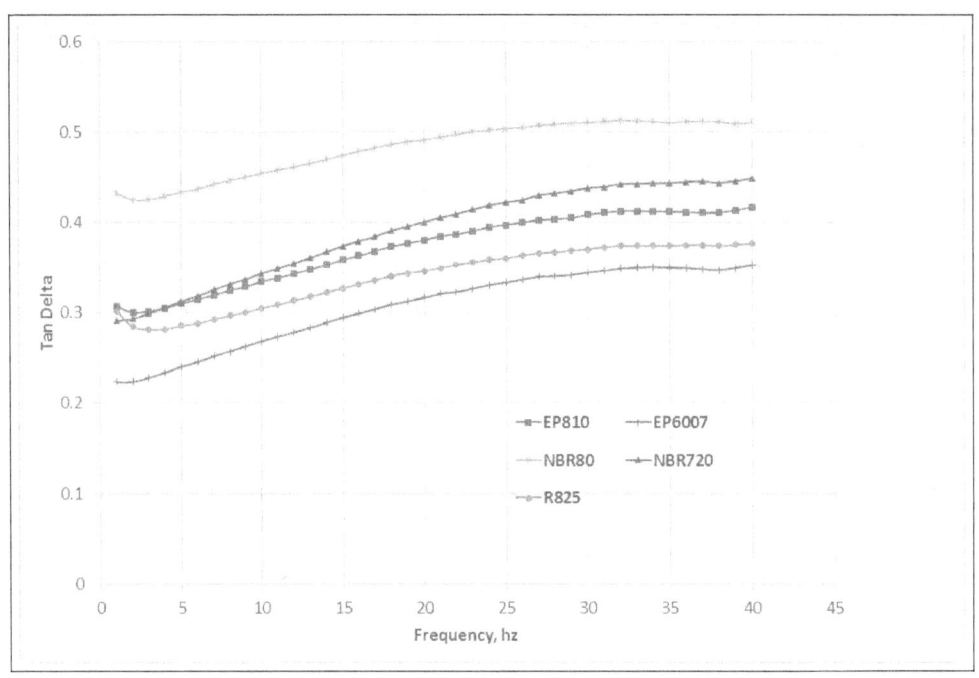

Figure .44: Plot of Tan delta Vs Frequency from a Frequency Sweep Test

The frequency sweep tests have been carried out by applying a pre-compression of 10 % and subsequently a displacement amplitude of 1 % has been applied in the positive and negative directions. Apart from tests on cylindrical and square block samples ASTM D5992 recommends the dual lap shear test specimen in rectangular, square and cylindrical shape specimens. Figure(.45) shows the double lap shear shapes recommended in the standard.

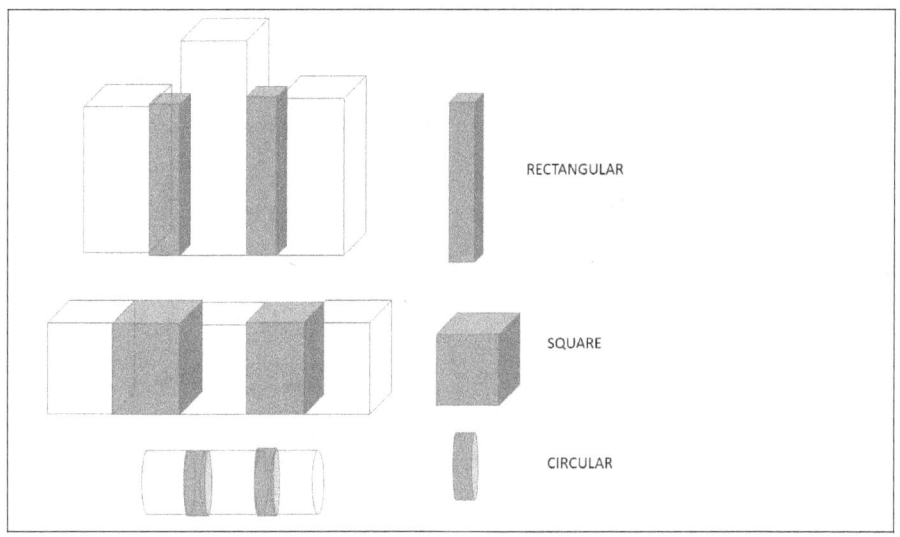

Figure .45: Double Lap Shear Shapes

# .16 References

1. Sperling, *Introduction to Physical Polymer Science*, Academic Press, 1994.

2. Ward et al., *Introduction to Mechanical Properties of Solid Polymers*, Wiley, 1993.

3. Seymour et al. *Introduction to Polymers*, Wiley, 1971.

4. Ferry, *Viscoelastic Properties of Polymers*, Wiley, 1980.

5. Goldman, *Prediction of Deformation Properties of Polymeric and Composite Materials*, ACS, 1994.

6. Menczel and Prime, *Thermal Analysis of Polymers*, Wiley, 2009.

7. Pete Petroff, *Rubber Energy Group Class Notes*, 2004.

8. ABAQUS Inc., *ABAQUS: Theory and Reference Manuals, ABAQUS Inc.*, RI, 02.

9. Attard, M.M., *Finite Strain: Isotropic Hyperelasticity*, International Journal of Solids and Structures, 2003.

10. Bathe, K. J., *Finite Element Procedures* Prentice-Hall, NJ, 96.

11. Bergstrom, J. S., and Boyce, M. C., *Mechanical Behavior of Particle Filled Elastomers*, Rubber Chemistry and Technology, Vol. 72, 2000.

12. Beatty, M.F., *Topics in Finite Elasticity: Hyperelasticity of Rubber, Elastomers and Biological Tissues with Examples*, Applied Mechanics Review, Vol. 40, No. 12, 1987.

13. Bischoff, J. E., Arruda, E. M., and Grosh, K., *A New Constitutive Model for the Compressibility of Elastomers at Finite Deformations*, Rubber Chemistry and Technology, Vol. 74, 2001.

14. Blatz, P. J., *Application of Finite Elasticity Theory to the Behavior of Rubberlike Materials*, Transactions of the Society of Rheology, Vol. 6, 1962.

15. Eberhard Meinecke, *Effect of Carbon-Black Loading and Crosslink Density on the Heat Build-Up in Elastomers*, Rubber Chemistry and Technology, Vol. 64, May 1991.

16. Boyce, M. C., and Arruda, E. M., *Constitutive Models of Rubber Elasticity: A Review*, Rubber Chemistry and Technology, Vol. 73, 2000.

17. Callister Jr., W. D., *Introduction to Materials Science and Engineering* John Wiley and Sons, NY, 1999.

18. Myers, M. A., , Chawla, K. K., *Mechanical Behavior of Materials* Prentice Hall, 1998.

19. Terrill., E. R. et al., *Modulus-Fatigue Resistance of Silica-filled Tire Tread Formulations* Fall 184th Technical Meeting of the ACS, Rubber Division, 2013.

20. Silva, A. J., Berry, N. G., Costa, M. F., *Structural and Thermo-mechanical Evaluation of Two Engineering Thermoplastic Polymers in Contact with Ethanol Fuel from Sugarcane* Material Research Vol.19 No.1, Jan/Feb 2016.

21. Menard, Kevin, P., *Dynamic-Mechanical Analysis*, CRC Press, Boca Raton, 1999.

22. Nielsen, L. E., *Mechanical Properties of Polymers*, Marcel Dekker, New York, 1974.

23. Tschloegl, N. W., *The Theory of Viscoelastic Behaviour*, Academic Press, New York, 1981.

24. Leyden, Jerry., *Failure Analysis in Elastomer Technology: Special Topics*, Rubber Division, 2003.

25. Baranwal, Krishna., *Elastomer Technology: Special Topics*, Rubber Division, 2003.

26. Srinivas, K., and Pannikottu, A., *Material Characterization and FEA of a Novel Compression Stress Relaxation Method to Evaluate Materials for Sealing Applications*, 28th Annual Dayton-Cincinnati Aerospace Science Symposium, March 2003.

27. Srinivas, K., *Systematic Experimental and Computational Mechanics Failure Analysis Methodologies for Polymer Components*, ARDL Technical Report, March 2008.

28. Dowling, N. E., *Mechanical Behavior of Materials, Engineering Methods for Deformation, Fracture and Fatigue* Prentice-Hall, NJ, 1999.

29. Hertz, D. L. Jr., *Designing with Elastomers*, Seals Eastern, Inc. 1983.

30. Srinivas, K., and Dharaiya, D., *Material And Rheological Characterization For Rapid Prototyping Of Elastomers Components*, American Chemical Society, Rubber Division, 170th Technical Meeting, Cincinnati, 2006.

31. Gent, N. A., *Engineering with Rubber: How to Design Rubber Components* Hanser Publishers, NY, 1992.

32. Horve, L., *Shaft Seals for Dynamic Applications*, Marcel Dekker, NY, 1996.

33. Pannikottu, A., Hendrickson, K. and Baranwal, K., *Comparison of Dynamic Testing Equipment and Test Methods for Tire Tread Compounds* ARDL Inc, OH, 1996.

34. Mars, W. V., and Fatemi, A., *A literature survey on fatigue analysis approaches for rubber*, International Journal of Fatigue, Vol. 24, 2002.

35. Belytschko T., Liu K.W, Moran B., *Nonlinear Finite Elements for Continua and Structures*, John Wiley and Sons Ltd, 2000.

36. Hasan O.A., Boyce M.C., *A constitutive model for the nonlinear viscoelastic viscoplastic bahavior of glassy polymers*, Polymer Engineering and Science 35, 1995.

37. Haward R.N., Young R.J., *The Physics of Glassy Polymers*, Chapman and Hall, 1997.

38. Simo J.C., *On a fully three-dimensional finite-strain viscoelastic damage model: formulation and computational aspects*, Computer Methods in Applied Mechanics and Engineering, Vol. 60, 1987.

39. Ogden, R. W. *Non-linear Elastic Deformations*, Dover Publications, New York, 1997.

40. Sasso, M., Palmieri, G., Chiappini, G., and Amodio, D. *Characterization of hyperelastic rubber-like materials by biaxial and uniaxial stretching tests based on optical methods*, Polymer Testing, Vol. 27, 2008.

41. Guo, Z., and L. J. Sluys, *Application of A New Constitutive Model for the Description of Rubber-Like Materials under Monotonic Loading*, International Journal of Solids and Structures, Vol. 43, 2006.

42. Kaliske, M., L. Nasdala, and H. Rothert, *On Damage Modeling for Elastic and Viscoelastic Materials at Large Strain. Computers and Structures*, Vol. 79, 2001.

43. T.G. Ebbott, R.L. Hohman, J.P. Jeusette, and V. Kerchman., *Tire temperature and rolling resistance prediction with Fnite element analysis*, Tire Science and Technology, Vol. 27, 1999.

44. KV Narasimha Rao, R Krishna Kumar, PC Bohara, and R Mukhopadhyay. *A Fnite element algorithm for the prediction of steady-state temperatures of rolling tires* Tire Science and Technology, Vol. 34(3), 2006.

45. Menczel., J.D., and Prime, R. B., *Thermal Analysis of Polymers*, Wiley, 2009.

46. Douglas Loy, *Class Lectures*, University of Arizona.

47. Joseph Padovan, Class Lectures, *Tire Design and Performance Prediction*, University of Akron.

48. Goldman, *Prediction of Deformation Properties of Polymeric and Composite Materials*, ACS, 1994.

49. Netzsch Instruments, *Class Notes and Machine Manuals*, 2004.

50. T A Instruments, *Class Notes and Machine Manuals*, 2006.

51. Perkin Elmer Industries, *Class Notes and Machine Manuals*, 2007.

52. Matsouka, *Relaxation Phenomena in Polymers*, Hanser, Frankfurt, 1993.

53. Danko, D. M., and Svarovsky, J. E., *An Application of Mini-Computers for the Determination of Elastomeric Damping Coefficients and Other Properties*, SAE Paper No. 730263.

54. Silberberg, Melvin., *Dynamic Mechanical Properties of Polymers: A Review*, Plus-Tech Equipment Corporation, Natick, Massachusetts, 1965.

55. Lakes, Roderick., *Viscoelastic Materials*, Cambridge University Press, 2009.

www.ingramcontent.com/pod-product-compliance
Lightning Source LLC
Chambersburg PA
CBHW062336220526
45469CB00008B/2731